WOODLICE AND WATERLICE
(ISOPODA: ONISCIDEA & ASELLOTA)
IN BRITAIN AND IRELAND

STEVE GREGORY
BRITISH MYRIAPOD AND ISOPOD GROUP

PHOTOGRAPHS BY
PAUL RICHARDS, DICK JONES AND THEODOOR HEIJERMAN

Published for: Biological Records Centre
                 NERC Centre for Ecology and Hydrology
                 Maclean Building
                 Benson Lane
                 Crowmarsh Gifford
                 Wallingford
                 Oxfordshire
                 OX10 8BB

By: FSC Publications
             Preston Montford Lane
             Shrewsbury
             SY4 1DU
             www.field-studies-council.org

ISBN: 978 0 9557672 8 9

Contents

Foreword

It is an unexpected pleasure to be asked to write a Foreword to this new account of the woodlice and waterlice of Britain and Ireland. When I first joined the Biological Records Centre in 1980, Paul Harding and Stephen Sutton were working on its predecessor, *Woodlice in Britain and Ireland* (1985), and Declan Doogue was finishing the remarkably thorough survey of Irish woodlice summarised in the *Distribution Atlas of Woodlice in Ireland* (Doogue & Harding 1982). I have been struck in reading this new text by how much has been added to our knowledge of the woodlice in the last 25 years by a well co-ordinated and dedicated group of recorders, many of them doubtless inspired by these earlier works. Paul Harding, who has done so much to encourage the recording of isopods in recent decades, appears, very appropriately, in a cameo role in this new book as author of the accounts of the four waterlice.

Steve Gregory has prepared the bulk of the text on behalf of the British Myriapod and Isopod Group. The accounts of the woodlice species give a vivid picture of them as living animals, with detailed accounts of their appearance, characteristic movements and rather precisely defined habitats, often illustrated by splendid photographs. It is not easy to distil the results of a wide-ranging national survey into a lively text, and Steve has doubtless succeeded so well because he brings to the task his own very extensive field experience.

Readers without specialist knowledge of the Isopoda will doubtless be struck by a range of similarities or contrasts with the groups they know themselves. Two points particularly interested me. It is clear from the text that students of woodlice are not prepared to accept the native/alien dichotomy as the sole criterion for the conservation of species, with native species treated (if threatened) as worthy of conservation action and alien species disregarded. Several species, such as *Acaeroplastes melanurus*, *Eluma caelatum* and the British population of *Oritoniscus flavus*, appear to be introductions and yet are valued members of our fauna. This parallels developments in flowering plants in the last decade, where we have realised that the long historical interaction between human populations and their environments has left us with a sequence of species introduced in ancient times, many of which have now become cherished plants. Another striking feature of the woodlice is the presence of numerous species with Atlantic ranges in Europe, such as *Ligia oceanica*, *Miktoniscus patiencei* and *Porcellionides cingendus*. This (if I dare say so) arguably makes the woodlice more important members of our fauna than superficially charismatic groups, such as the butterflies, which are too thermophilous to be well-represented in our oceanic islands.

This book shows how even a relatively small band of recorders can not only contribute to the knowledge of their own specialist group but also produce a text which illuminates for the rest of us wider aspects of the biogeography and conservation of British and Irish biodiversity. It is a particularly appropriate volume to launch a planned series of atlases to be published in partnership with the Field Studies Council, an arrangement which we hope will bring the publications of the biological recording schemes to a wide audience.

C. D. Preston,
Biological Records Centre, June 2009

Acknowledgements

Since the publication of *Woodlice in Britain and Ireland* (Harding & Sutton, 1985) the Non-marine Isopod Recording Scheme has been run by a succession of enthusiastic and capable organisers who have willingly given their time and expertise to encourage others. George Fussey, the late Steve Hopkin and David Bilton have been instrumental in co-ordinating the collection and collation of woodlice and waterlice records upon which this updated publication is based. Personally, I am indebted to Steve Hopkin for his support and encouragement when, as a complete novice, I began to study these interesting animals in the late 1980s.

This publication would not have been possible without the tireless work of Paul Harding, both in his former capacity as head of Biological Records Centre and as a stalwart of the British Isopod Study Group and its successor, the British Myriapod and Isopod Group. I am indebted to Paul for writing the draft species accounts for the Asellidae, shamefully a group I know little about.

I am extremely grateful to Glyn Collis for dealing with my numerous queries and for providing much up-to-date literature. Paul Harding, Jon Daws and John Harper have also provided detailed and stimulating discussion on a wide variety of topics. Paul Richards (www.invertebrate-images.co.uk), Dick Jones and Theodoor Heijerman have all generously allowed the reproduction of their photographs, free of charge.

Many other BMIG members have provided useful feedback on various versions of draft text and species distribution maps and/or have supplied copies of relevant literature. My thanks go to Keith Alexander, Roy Anderson, David Bilton, Matty Berg, Martin Cawley, Gordon Corbet, Michael Davidson, Adrian Fowles, Ian Morgan, Eric Philp and Paul Richards, and anyone else who has helped in any way with the production of this volume. Their efforts have undoubtedly helped improve the accuracy and usefulness of this work.

The guidance and assistance provided by staff at BRC is acknowledged. Val Burton has inputted much of the data from record cards. Henry Arnold, Helen Roy and Stephanie Ames have been responsible for uploading species records, running automated validation procedures and plotting species maps. I also thank them for their quick response to my frequent queries and requests. I am grateful to Chris Preston for his constructive comments on the text. The maps have been prepared using DMAP mapping software developed by Dr Alan J. Morton.

The production of this distribution atlas would not have been possible without the huge contribution made by the many active recorders located throughout Britain and Ireland. A number of individuals and organisations have willingly contributed substantial datasets. Roy Anderson and Martin Cawley readily made available their respective datasets for Ireland. Peter Harvey provided access to the huge wealth of information held by the Essex Field Club (www.essexfieldclub.org.uk). The Environment Agency kindly permitted the use of their waterlice records within this publication.

Over 1,000 individuals, some long dead, and organisations have submitted records to the Non-marine Isopod Recording Scheme. Those that have submitted more than 10 records to the scheme are listed below.

Abbott, A.M.	Allenby, K.G.	Bagnall, R.S.	Bilton, D.T.
Addey, J.E.	Anderson, R.	Barber, A.D.	Bishop, J.D.
Al-Dabbagh, K.Y.	Arnold, J.	Barnes, S.	Bishop, M.H.
Alexander, K.N.A.	Askins, M.C.	Bates, J.	Bloor, K.P.
Allen, Louise	d'Ayala, R.L.	Bentley, D.	Blower, J.G.

Bolton, D.
Boot, K.
Bowler, N.
Boyce, D.
Boyd, J.
Boyd, J.M.
Bratton, J.
Bray, R.P.
Britt, D.P.
Bullard, E.
Burns, P.F.
Byford, T.E.
Cameron, R.A.D.
Campbell, J.M.
Cave Research Group
Cawley, M.
Chapman, R.A.
Chater, A.O.
Chatfield, J.E.
Cheek, T.M.
Cheesman, C.
Christie, L.
Clements, D.K.
Clements, H.A.B.
Clinging, R.
Collinge, W.E.
Collis, G.M. & V.D.
Colston, A.
Cope, S.
Copson, P.J.
Corbet, G.B.
Cotton, D.C.F.
Crawford, A.K.
Crawshaw, D.I.
Crittenden, M.
Crosby, T.S.
Crowson, R.A.
Cubbon, B.D.
Daniel, R.
Darlington, J.P.E.C.
Davey, S.
David, C.
Davidson, M.B.
Davies, L.
Davis, B.N.K.

Davis, R.C.
Daws, J.
Dawson, N.
Denton, J.
Denton, M.
Dixon, J.
Dolling, M.
Doogue, D.
Drake, C.M.
Driscoll, R.J.
Duffey, E.
Dupey, Dr.
Eeles, J.
Ellis, A.E.
Ellis, E.A.
Ely, W.A.
Environment Agency
Evans, I.
Evans, I.M.
Evans, L.
Evans, M.S.
Eversham, B.
Eyres, M.
Fairhurst, C.P. & J.
Farmer, G.
Farrell, L.
Felton, C.
Fleming, J.
Fleming, S.
Fogan, M.
Foster, G.N.
Foster, N.H.
Fowler, A.P.
Frankel, B.
Fraser, A.J.L.
Fryer, G.
Funnell, D.
Fussey, G.D.
Garland, S.P.
Garrad, L.
George, R.S.
Gillham, A.
Gilson, H.C.
Gledhill, T.
Glennie, E.A.

Goddard, D.G.
Godfrey, A.
Goldie-Smith, E.K.
Goldsmith, J.G.
Green, G.H.
Gregory, S.J.
Grove, S.J.
Guntrip, D.
Guyoncourt, D.
Hadley, M.
Hames, C.A.C.
Hamilton, J.D.
Hammond, P.
Hancock, E.G.
Hand, S.C.
Hanson, D.
Harding, P.T.
Harper, J.F.
Harris, G.J.
Hart, A.
Harvey, M.C.
Harvey, N.
Harvey, P.R.
Hassall, M.
Hawkins, K.
Heaver, D.
Hennessy, M.
Higgins, D.
Hill, K.
Hills, C.C.
Holdich, D.M.
Holland, D.G.
Holloway, R.
Hopkin, S.P.
Hopkins, I.J.
Howard, H.W.
Howe, M.A.
Hughes, M.R.
Hunnisett, J.
Hunter, M.
Hynes, H.B.N.
Ing, B.
Irwin, A.G.
Ivemey-Cook, P.
Jackson, N.C.S.

Jagoe, R.B.
Johnson, A.C.
Jones, G.H.
Jones, P.E.
Jones, R.A.
Jones, R.E.
Jones, S.P.
Kearns, H.
Keay, A.N.
Kendall, P.
Key, R.S.
Kime, R.D.
Kingley, M.R.
Kirby, P.
Lambert, D.
Langston, M.R.
Larwood, H.J.
Law, N.
Lawrence, P.N.
Lee, P.
Leicester
  Underwater
  Exploration Club
Lewenz, P.
Lewis, J.W.
Line, J.M.
Lloyd, O.C.
Lloyd-Evans, L.
Lovering, T.A.
Loxton, R.G.
Loynds, G.A.
Luff, M.L.
Lynch, D.
Macan, T.T.
Mackay, J.
Mander, P.B.
Mann, D.J.
Masher, G.
Maughan, E.
Maxwell, W.G.R.
McAllister, R.I.
McCutcheon, D.E.
McDonald, R.
McGrath, D.
McWilliam, S.J.

Meiklejohn, J.W.
Meloy, B.J.
Messer, D.
Messer, L.
Metcalfe, R.J.A.
Miles, P.M.
Millatt, W.E.
Mitchell, D.W.
Mitchell, H.
Moller, G.J.
Monteith, C.
Moon, H.P.
Morgan, I.K.
Morgan, M.J.
Moriarty, C.
Morris, M.G.
Morris, R.
Morton, A.
Moseley, C.M.
Mothersill, C.
Murphy, C.M.
Murrell, S.G.
Musson, D.
Nau, B.S.
Newman, N.A.
Newton, J.M.
Nisbett, G.
Norledge, R.
Norman, A.M.
Norris, A.
O'Connor, J.P.
Oliver, P.G.
Olsen, K.M.
O'Mahoney, P.
O'Meara, M.
Osborn, A.G.H.
Osley, N.J.
Oxford, G.
Pack Beresford, D.R.
Palmer, M.
Parsons, M
Parton, P.J.
Partridge, J.
Passmore Edwards
  Museum

Paton, C.I.
Paul, C.R.C.
Pearce, E.J.
Phillips, A.
Phillips, R.A.
Philp, B.
Philp, E.G.
Plant, C.W.
Plant, R.A.
Powell, R.P.
Preston, C.D.
Price, R.
Quick, H.E.
Rands, E.B.
Rapp, W.F.
Rawcliffe, C.P.
Read, H.J.
Read, R.W.J.
Reardon, N.M.
Redgate, N.
Redshaw, E.J.
Rees, C.J.
Reid, A.
Reid, C.
Relton, J.
Reynolds, P.
Richards, J.P.
Richardson, D.T.
Riley, T.H.
Robinson, N.A.
Roper, P.
Rothwell, D.
Ruffell, R.
Rundle, A.J.
Rutherford, M.
Sage, J.R.
Sander. J.I.
Scotter, C.
Scott-Langley, D.
Seaward, D.R.
Sharrock, J.T.R.
Shotton, M.F.
Side, K.C.
Simpson, A.
Skidmore, P.

Slawson, G.C.
Smith, C.J.
Smith, M.N.
Southward, E.C.
Spenser, G.G.
Spirit, M.G.
Standen, V.
Stark, L.
Stebbings, R.E.
Stelfox, A.W.
Stoke-on-Trent
  Environmental
  Survey Team
Stott, N.
Stott, W.G.R.
Suffolk Wildlife Trust
Sumner, A.
Sunderland, K.D.
Sutcliffe, D.W.
Sutcliffe, R.
Sutton, C.A.
Sutton, S.L.
Teagle, W.G.
Telfer, M.G.
Tew, G.S.
Thomas, T.J.
Tinning, P.C.
Topley, P.
Townsend, M.C.
Trett, J.
Trew, A.
Turnbull, S.
Turner, M.A.
Turner, S.
Tynen, M.
Varndell, I.
Wade, P.M.
Walden, H.
Walker, C.
Walker, J.
Wallace, I.
Wallace, I.D.
Walters, M.G.
Ward, L.K.
Ward, P.A.

Wardaugh, M.
Warmingham, S.C.
Warnact Wetland
  Group
Warne, A.C.
Warren, D.M.E.
Warwick, T.
Watson, R.A.
Watson, R.G.
Watson, W.
Webb, W.M.
Welch, R.
Welch, R.C.
Welsh Peatland
  Invertebrate
  Survey
Welsh Water
  Authority
Weston, M.R.
Westwood, W.D.
White, A.J.
Whitehead, P.F.
Whiteley, D.
Williams, W.D.
Willows, R.I.
Wills, H.J.
Wilson, F.M.
Winnal, R.
Wright, I.R.
Wright, J.
Wright, R.
Yates, B.J.
Youell, S.

Introduction

Michael Usher in the foreword to *Woodlice in Britain and Ireland* (Harding & Sutton, 1985) noted that a lot is known about a few species, and little is known about a lot of species. Although to some extent this remains true, the publication of 'WIBI' succeeded in stimulating considerable interest in the recording of woodlice throughout Britain and Ireland. As a direct result, we are now in a much better position to understand the correlation between the distribution of species and environmental factors. The habitat preferences of scarce woodlice such as *Armadillidium pictum*, *Porcellionides cingendus* and *Halophiloscia couchii* are now much better understood. As Michael Usher hoped, WIBI did indeed go out of date very quickly and this updated account of the woodlice in Britain and Ireland is long overdue. The aquatic waterlice (or water-slaters), a close relative to the woodlice, are also included within this volume.

Woodlice and waterlice are arthropods and in common with other arthropod groups, such as millipedes, centipedes, spiders and insects, they have a segmented body bearing jointed legs and a hard exoskeleton. More specifically they are Crustaceans, related to crabs, shrimps and lobsters and within the Crustacea they belong to the Order Isopoda. The isopod body is characteristically flattened from top to bottom, facilitating walking, and is differentiated into three sections: the head, the pereon (thorax) and the posterior section, the pleon (abdomen). The pereon has seven segments (pereonites), each bearing a pair of legs (pereopods) which are similar to one another (hence Isopoda). Isopods differ from other Crustacean groups in lacking gills concealed beneath a carapace, but instead breathe using specialised lamellar gill-like pleopods beneath the pleon. In some terrestrial woodlice the pleopods bear respiratory structures known as 'pleopodal lungs' that may be seen as white patches. Our inland (non-marine) isopods are divided into two sub-orders: Asellota, the aquatic waterlice, and Oniscidea, the terrestrial woodlice. They range in size from 2mm to 30mm.

Waterlice

The Waterlice (Isopoda: Asellota), as their name suggests, are aquatic species occurring in a range of freshwater habitats, including marshes, streams, ponds and lakes. Four species, all within the family Asellidae, are included on the British list, although only two of these species occur in Ireland. The ubiquitous *Asellus aquaticus*, and its close relative *Proasellus meridianus*, will be familiar to anyone who has dipped a net into a pond. One species, *Proasellus cavaticus*, is exclusively subterranean, occurring deep underground in caves or aquifers. *Caecidotea communis* is an introduction from North America.

The phyletic relationships of Asellidae genera and species have been the subject of much debate and research, including in recent years RNA and DNA sequencing. According to authors such as Hidding, Michel, Natyaganova, Sherbakov & Yu (2003) and Verovnik, Sket & Trontelj (2005) the genus *Asellus* has a predominantly east Palaearctic distribution, but *Proasellus* seems to be primarily Mediterranean. *Caecidotea* is North American and quite distinct from Palaearctic genera. Both these publications, and many others, describe how endemic speciation commonly occurs in geographically isolated populations of asellids and other freshwater Malacostraca. This is particularly noticeable in Lake Baikal in Siberia, and in subterranean waters of karst areas in the Balkans.

A. aquaticus and *P. meridianus* have been the subjects of much ecological and physiological research. They are important elements of the biota of many freshwater systems, particularly in still and slow-flowing waters. They feed upon a variety of organic detritus. Algae and fungi have been recorded as important sources of food (Moore, 1975; Rossi & Fano, 1979; Graça, Maltby & Calow, 1993). In a study of *A. aquaticus*, Rossi and Vitagliano-Tadini (1978) noted that hyphomycetes in the faeces of adults increased the survival and growth of juveniles.

Waterlice form part of the food resource for other invertebrates, such as dragonfly and water beetle larvae, and for vertebrates such as fishes, newts and dippers. *A. aquaticus* and *P. meridianus* have been identified as intermediary hosts for some parasites of fishes, such as larval flukes. The ability of asellids to accumulate high concentrations of some heavy metals (Martin & Holdich, 1986) may implicate them in the movement of these potentially lethal pollutants further up the food-chain.

Woodlice

Woodlice (Isopoda: Oniscidea) are terrestrial forms, being the most successful group of Crustaceans to have colonised dry land. Schmalfuss (2004) recognised around 3,500 valid species worldwide, but many more undoubtedly await discovery. It is thought that this successful transition to land was made via the seashore and several species remain confined to the supralittoral zone. In many ways woodlice ancestors were pre-adapted to invade dry land. The dorso-ventrally flattened body facilitates walking and the brood pouch (marsupium) beneath the body provides a moist environment for incubating eggs and young (Hopkin 1991a). However, their ultimate success on dry land can be attributed to a number of ecophysiological adaptations that overcome potential problems of oxygen uptake, water regulation, nitrogen discharge and tolerance to toxic substances (Wieser, 1984).

In Britain forty species of woodlice have been recorded outdoors; thirty-one in Ireland. Artificial climates, such as those maintained inside glasshouses, provide suitable conditions for an additional twelve introduced 'alien' species. Four ubiquitous species, *Oniscus asellus*, *Porcellio scaber*, *Philoscia muscorum* and *Trichoniscus pusillus* agg., are found in large numbers across much of Britain and Ireland, whilst the pill-woodlouse, *Armadillidium vulgare*, can be abundant in the south and east. These are the familiar 'woodlice' generally encountered by naturalists and the public alike. Many of the less common species live unnoticed in our countryside and towns, typically inhabiting damp dark places, such as under logs and stones or among moss and leaf-litter or deep within soil crevices. A few, such as *A. vulgare*, are able to tolerate quite dry conditions and may be active during the day.

Woodlice have been placed into a number of 'body-construction categories' correlated to their ecological and behavioural strategies (Schmalfuss, 1984). Perhaps most familiar are the 'rollers', typified by the pill-woodlouse *A. vulgare*, that roll into a ball when provoked by a potential predator. When disturbed 'clingers' such as *O. asellus* and *P. scaber* remain motionless with their broad tergites tightly clamped to the underlying substrate to prevent them from being easily dislodged. *P. muscorum* is a good example of a 'runner' able to use its narrow body and long legs to escape at speed when disturbed. Many tiny trichoniscids, such as *Haplophthalmus* spp. and *Trichoniscoides* spp., are 'creepers'. These tend to have short weak legs for negotiating narrow soil crevices and their tuberculate or ridged dorsal surfaces prevents adhesion to waterlogged surfaces. As with all strategies there are 'non-conformists', such as *Platyarthrus hoffmannseggii*, which inhabits ants nests and does not fit easily into any of the above categories.

Figure 1. *Platyarthrus hoffmannseggii* and *Lasius flavus*. ©Paul Richards.

Woodlice are a key component in the processes of decomposition and nutrient recycling. Depending on the species, they consume a wide variety of material, including lichen, algae, fungi, partly decomposed plant material and even flesh from dead animals. Woodlice also consume a proportion of their own faecal pellets. This is essential because their digestive system is very inefficient and much of whatever they consume, including vital nutrients needed for survival, passes straight through the gut. They rely upon bacterial activity prior to reingestion to unlock nutrients, such as copper needed to make haemocyanin for their blood, within the partly digested faecal pellets (Sutton, 1972). The physical acts of chewing organic matter, mixing leaf litter and dispersing fungal spores means that woodlice speed up the recycling of nutrients back into the soil and ultimately back into the growth of new vegetation.

Woodlice are a dominant component of the soil arthropod fauna and their habit of browsing dead organic material, and any pollutants contained within it, make them potential environmental indicators. They are sensitive to many residual chemicals, such as pesticides, and marked differences in abundance may be observed between conventional and organic farming systems (Paoletti & Hassall, 1999). Some species, notably *P. scaber*, are tolerant of heavy metals, including lead, copper, zinc and cadmium. Large concentrations of these metals can be stored harmlessly within the body and correlate precisely with levels of contamination within their environment (Hopkin, Hardisty & Martin, 1986; Hopkin, 1989). Thus, certain species of woodlice make useful bioindicators of heavy metal pollution, such as that resulting from industrial processes.

Woodlice have many natural enemies. The woodlouse spider, *Dysdera crocata* and its smaller cousin *D. erythrina*, are specialist woodlice predators. They have modified chelicerae adapted to pierce the body of adult woodlice and the latter species has been shown to reject most other potential prey items (Řezáč, Pekár & Lubin, 2008). Generalist predatory invertebrates, including lycosid spiders, centipedes and beetles, are known to consume woodlice in the wild. Adult woodlice are generally well armoured and are able to secrete repellent chemicals, so usually it is immature individuals that are taken (Sunderland & Sutton, 1980). In some cases the corpses of dead woodlice have been scavenged. Vertebrates, such as toads and shrews, will also feed upon woodlice.

Figure 2. *Dysdera crocata* and *Oniscus asellus*. ©Paul Richards.

Rhinophorid flies (Diptera: Rhinophoridae) are specialist parasitoids of woodlice. The fly's larva consumes its woodlouse host from within. The woodlouse is kept alive until the vital organs are finally eaten prior to the larva pupating within the host's empty exoskeleton. There are just seven species of these small black bristly flies in the British Isles. Each tends to favour one species of woodlouse, often within a particular type of habitat (Sutton, 1972), but recent work in the Netherlands suggests that Rhinophorid flies may compete for the same hosts within similar microhabitats (Wijnhoven, 2001b).

A number of pathogens infect woodlice, some having interesting effects on the host. Iridovirus can infect many different species of woodlice (Wijnhoven & Berg, 1999). In time a characteristic bright bluish-purple colour develops within the infected specimen as crystalline structures accumulate within the diseased body tissues shortly before death. In Britain *T. pusillus* agg. seems to be the most frequently encountered species with iridovirus infection. *Wolbachia* bacteria also occur widely in woodlice (Bouchon, Rigaud & Juchault, 1998). The bacteria are passed on to the next generation through maternal transmission (ova). In order to maximise its own reproductive potential *Wolbachia* turns male woodlice into functional females by disrupting their sexual hormones.

As described by the late Steve Hopkin (Hopkin, 1991a), woodlice are fascinating and endearing creatures, and much remains to be discovered about them, not just on their distribution and biology, but also their physiology, their use in ecotoxicology, their interactions with microbiota and much more.

Recording in Britain and Ireland

History of woodlice recording

This account summarises progress made in the recording of native and naturalised woodlice species in Britain and Ireland. It highlights the discovery of each species and some of the key personalities involved. Alien species associated with heated glasshouses are considered separately.

The early years: 1830-1898

The first published British records were made by Leach in 1830, who recorded six species, *Armadillidium vulgare*, *Ligia oceanica*, *Oniscus asellus*, *Philoscia muscorum*, *Porcellio laevis* and *P. scaber*. In 1836 Templeton reported five of these species from Ireland, with Thompson adding *P. muscorum* in 1856. The following year Kinahan added six species to the Irish list, including the discovery of *Porcellionides cingendus* new to science. The additional species were: *Oritoniscus flavus*, *Porcellionides pruinosus*, *Porcellio dilatatus*, *Porcellio spinicornis* and *Trichoniscus pusillus*. In Britain, Bate & Westwood added *Androniscus dentiger*, *Cylisticus convexus* and *Halophiloscia couchii* in 1868. In 1873 Stebbing added *Ligidium hypnorum* from Britain. It is apparent from Fig. 3 that the number of woodlice known from Britain and Ireland increased steadily throughout the 19th Century and by 1873 stood at seventeen species.

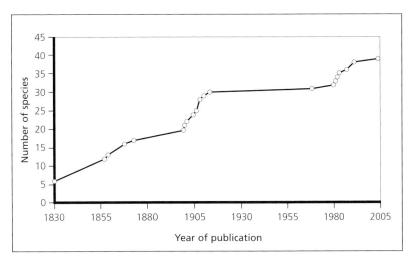

Figure 3. The chronological growth of the British and Irish list of native and naturalised woodlice species.

A period of great activity: 1899 to 1913

It was 26 years before another species was added to the British and Irish check-list, but the beginning of the 20th Century saw a period of considerable activity. In 1899 Norman added three species, *Armadillidium depressum*, *A. nasatum* and *Haplophthalmus danicus*, and made the first discovery of *P. cingendus* on the British mainland. At the turn of the century Kane and Scharff discovered *Haplophthalmus mengii* and *Armadillidium pulchellum*, respectively, in Ireland. Soon after (1904) Norman & Brady found these two species in Britain and added *Trachelipus rathkii* and *Trichoniscoides albidus* to the overall check-list. Between 1906 and 1908 Bagnall added *Armadillidium album* and *Trichoniscus pygmaeus* and Patience added *Trichoniscoides sarsi* (all from

Britain), whilst in Ireland Pack Beresford added *Eluma caelatum* and *Halophiloscia couchii*. This busy period closed with the discovery by Scharff of *Acaeroplastes melanurus* in Ireland (1910) and by Standen of *Armadillidium pictum* in Britain (1913). The culmination of this effort was the discovery of thirteen additional native and naturalised woodlice species, bringing the final total for this period to thirty.

The 'dark ages': 1914 to 1967

This period is easily dismissed since no additional species were added for half a century and Harding (1977) shows that the reliability of many records published in the early part of this period, particularly those attributable to W.E. Collinge, is questionable. However, this period saw the publication of the first *Synopsis of British Woodlice* by Edney (1954). This was the first workable key to all British and Irish species, including glasshouse aliens, that was available to British workers and represented a major step forward to the recording of woodlice. Had this work not been available, the surveys initiated by the British Isopod Study Group (BISG) in the late 1960s would have been a considerably more daunting task.

The rise of BISG: 1968 to 1985

1968 saw the inception of the British Isopod Survey Scheme and, coincidently, the first addition to the British and Irish species list since 1913 with the discovery by Sheppard (1968) of *Trichoniscoides saeroeensis*. The following year, under the leadership of Stephen Sutton, John Metcalfe and Paul Harding, the British Isopod Study Group was formed. A revised and more accessible identification key was included within Sutton's *Woodlice* (1972). Fieldwork stimulated by the publication of the *Provisional Atlas* (Harding, 1976a) resulted in the discovery of four additional species in the early 1980s. These were *Stenophiloscia glarearum* (*S. zosterae*) (Harding, Cotton & Rundle, 1980), *Metatrichoniscoides celticus* described new to science (Oliver & Trew, 1981), *Miktoniscus patiencei* (Oliver & Sutton, 1982) and *Buddelundiella cataractae* (Oliver, 1983). These increased the list to thirty-five species. Throughout this period the Non-marine Isopod Recording Scheme remained in the capable hands of Paul Harding, until handing over to George Fussey in 1982. In 1985 the much anticipated *Woodlice in Britain and Ireland* (Harding & Sutton, 1985) was published.

After Woodlice in Britain and Ireland: 1986 to 2007

The publication of *Woodlice in Britain and Ireland* maintained the impetus in woodlice recording, but additional species were not easily found. In 1985 Steve Hopkin took over the recording scheme and promptly added two cryptic species, *Haplophthalmus montivagus* and *Trichoniscoides helveticus* (Hopkin & Roberts, 1987; Hopkin, 1990b). A third species, *Metatrichoniscoides leydigii*, was added by Steve Gregory (Hopkin, 1990a). A species of *Chaetophiloscia* found by Jones and Pratley (1987) to be naturalised on the Isles of Scilly remains unidentified. David Bilton ran the scheme from 1991 to 1999 and came close to adding a species new to science with his description of the subspecies *O. asellus* ssp. *occidentalis* (Bilton, 1994). Also in 1994, *O. flavus* was recorded from Britain for the first time by Ian Morgan (1994). Schmalfuss (2004) elevated the two forms of *T. pusillus* aggregate to full species, therefore adding *Trichoniscus provisorius* to our check-list. In 2000 BISG merged with the British Myriapod Group to form the British Myriapod and Isopod Group (BMIG) and the recording scheme was transferred to Steve Gregory. The number of native or naturalised woodlice species currently known in Britain and Ireland is forty. Interest in woodlice recording continues at a high level, supported by the publication of Steve Hopkin's very accessible AIDGAP key (Hopkin, 1991a) and by a new Synopsis (Oliver & Meechan, 1993).

Glasshouse aliens

Collecting from heated glasshouses, such as those at the botanic gardens of Kew, Glasgow, Edinburgh or Dublin, was popular in the early 20th Century. Between 1908 and 1918 eight glasshouse aliens were recorded by Patience, Bagnall, Jackson and Collinge. Four, *Miktoniscus linearis*, *Cordioniscus stebbingi*, *Styloniscus spinosus* and *Setaphora patiencei*, were described new to science. The remainder were *Trichorhina tomentosa*, *Agabiformius lentus*, *Nagurus cristatus* and *Nagurus nanus*. In 1947, Holthuis added two further aliens from Kew Gardens; *Reductoniscus costulatus* and *Burmoniscus meeusei* (the latter also described new to science). In recent years collecting in glasshouses has not been popular. It was not until 1987 that *Styloniscus mauritiensis* was added to the list of alien woodlice (Collis & Harding, 2007) and, more recently, *Venezillo parvus* in 2005 (Gregory, 2009). The number of known alien species now stands at twelve, but it is probable that other species await discovery.

British Myriapod and Isopod Group

The British Myriapod and Isopod Group (BMIG) aims to promote the study of myriapods (millipedes, centipedes, pauropods and symphylans) and non-marine isopods (woodlice and waterlice) by:

- providing up-to-date information
- supporting members
- holding field meetings.

It is affiliated to the British Entomological and Natural History Society (BENHS). Membership currently stands at about 250 individuals, of whom a fifth are overseas members. Members of BMIG have played leading roles in international organisations and continue to do so today. The Non-marine Isopod Recording Scheme (in parallel to the millipede and centipede schemes) is run under the auspices of BMIG.

Members receive a copy of the Group's newsletter twice each year, which includes an update of recording scheme activity and other topics, such as useful hints on collecting techniques or species identification. BMIG also publishes a peer-reviewed Bulletin (ISSN 1475-1739), usually annually. The main field meeting is held just before or just after Easter, and involves a residential weekend, usually in an under-recorded area of Britain or Ireland. The field meeting is attended by a mixture of beginners, non-specialists and specialists and provides the opportunity to undertake fieldwork, to discuss common interests and to learn from one another. In some years additional meetings are arranged and these have included overseas trips to Hungary and to northern Spain. New members are most welcome. Further details of BMIG and its activities are available from the website at www.bmig.org.uk.

Non-marine Isopod Recording Scheme

Background to the scheme

The initial Isopod Recording Scheme was established in 1968 under the auspices of the former British Isopod Study Group and was intended to cover all British and Irish species of Isopoda, including free-living and parasitic marine species. In 1970 the marine species were split off to form an independent recording scheme (that regrettably ceased to operate after only a few years). This left the Non-marine Isopod Recording Scheme to cover the terrestrial woodlice (Oniscidea) and the aquatic waterlice (Asellidae). A detailed account of the historical background and development of the survey scheme is given in Harding & Sutton (1985). It might perhaps be more logical to record Asellidae with other freshwater Malacostraca, such as Amphipoda, but hitherto there is no recording scheme for freshwater Amphipoda, other than for subterranean species (Knight, undated [2007]).

In 1969 discussions with organisers of the nascent millipede and centipede recording schemes led to the adoption of a habitat classification scheme, designed primarily for terrestrial habitats, that pioneered an ecological approach to species recording. Instead of simply aiming to collect information for mapping species distribution on a 10km square basis, the record cards included a detailed hierarchical habitat classification, including microsite information. The analysis of this habitat data was instrumental to the success of *Woodlice in Britain and Ireland* (WIBI) (Harding & Sutton, 1985). This work did not include the aquatic Asellidae, mainly because most of the available records had been summarised by Moon and Harding (1981).

Species recording

The Non-marine Isopod Recording Scheme remains active and recent work, culminating in the production of this current book, has concentrated on improving our understanding of the distribution and biology of our species. The scheme would welcome more records for all species. Many of the records plotted in this volume, particularly in Ireland and Scotland, are twenty-five years old and are in urgent need of updating. However, all records, even if they do not add new dots to the distribution maps, are valuable. New contributors to the scheme are most welcome and support and guidance will be offered where required.

Where records are submitted for a species that is difficult to identify (see 'critical species' section, page 155) or where the record seems surprising based on our current understanding of that species' distribution and biology the scheme organiser may ask to see a voucher specimen. The advantage of this approach is that the accuracy of species' records held by the scheme is ensured and that a small group of reliable recorders, supported by the central scheme organiser, is created.

An updated non-marine isopod recording card (RA84) has recently been produced (see Fig. 48, page 157) and is available from BRC. Guidelines on how to submit biological records are included within the section on collecting and recording (page 153). Requests for assistance with identifications and submission of records of woodlice and waterlice should be sent to:

Steve Gregory
c/o Northmoor Trust, Hill Farm, Little Wittenham, Abingdon, Oxfordshire, OX14 4QZ. UK.
E-mail: steve.gregory@northmoortrust.co.uk

Helen Roy
Biological Records Centre, CEH Wallingford, Crowmarsh Gifford, Oxfordshire, OX10 8BB UK.
E-mail: hele@ceh.ac.uk

Summary of the dataset

Tables 1 and 2 indicate the number of records for asellids and oniscids, respectively, received by the Non-marine Isopod Recording Scheme up to the end of December 2007. Species are ranked according the total number of records submitted. The number of records submitted by the most prolific recorders is shown in Table 3.

Species aggregates (agg.) are groups of closely related species that can only be separated by microscopic examination of a male specimen. If a specimen cannot be referred to the precise species then the record can only be submitted for the species aggregate. These are:

- *Haplophthalmus mengii* agg. comprising *H. mengii* (Zaddach) & *H. montivagus* Verhoeff
- *Trichoniscoides sarsi* agg. comprising *T. sarsi* Patience & *T. helveticus* (Carl)
- *Trichoniscus pusillus* agg. comprising *Trichoniscus pusillus* Brandt & *T. provisorius* Racovitza

Few recorders examine males of *T. pusillus* agg., hence the large number of records for this ubiquitous species aggregate.

Table 1. Number of waterlice records received up to December 2007.

Taxon	No. records	% Total	Rank
Asellus aquaticus	62,866	90.28	1
Proasellus cavaticus	103	0.15	3
Proasellus meridianus	6,659	9.56	2
Caecidotea communis	5	0.01	4
Total number of waterlice records	69,633	100 %	

Table 2. Number of woodlice records received up to December 2007.

Taxon	No. records	% Total	Rank
Ligia oceanica	1,831	2.13	9
Ligidium hypnorum	616	0.72	16
Androniscus dentiger	2,598	3.02	6
Buddelundiella cataractae	17	0.02	= 38
Haplophthalmus danicus	1,241	1.44	10
Haplophthalmus mengii agg.	769	0.90	13
Haplophthalmus mengii	494	0.58	19
Haplophthalmus montivagus	40	0.05	34
Metatrichoniscoides celticus	17	0.02	= 38
Metatrichoniscoides leydigii	1	< 0.01	42
Miktoniscus patiencei	83	0.10	32
Oritoniscus flavus	155	0.18	28
Trichoniscoides albidus	332	0.39	21
Trichoniscoides helveticus	26	0.03	36

Continued overleaf

Table 2 (continued).

Taxon	No. records	% Total	Rank
Trichoniscoides saeroeensis	288	0.34	23
*Trichoniscoides sarsi agg.	38	0.04	35
Trichoniscoides sarsi	22	0.03	37
Trichoniscus provisorius	183	0.21	25
*Trichoniscus pusillus agg.	11,942	13.90	4
Trichoniscus pusillus	161	0.19	27
Trichoniscus pygmaeus	2,187	2.55	8
Halophiloscia couchii	136	0.16	29
Stenophiloscia glarearum	11	0.01	40
Philoscia muscorum	12,521	14.57	3
Oniscus asellus (all taxa)	18,464	21.49	1
Oniscus asellus ssp. *asellus*	149	-	-
Oniscus asellus ssp. *occidentalis*	52	-	-
Oniscus asellus x *occidentalis*	34	-	-
Platyarthrus hoffmannseggii	2,334	2.72	7
Armadillidium album	167	0.19	26
Armadillidium depressum	598	0.70	17
Armadillidium nasatum	760	0.88	14
Armadillidium pictum	44	0.05	33
Armadillidium pulchellum	331	0.39	22
Armadillidium vulgare	7,216	8.40	5
Eluma caelatum	97	0.11	30
Cylisticus convexus	510	0.59	18
Porcellio dilatatus	266	0.31	24
Porcellio laevis	87	0.10	31
Porcellio scaber	16,078	18.71	2
Porcellio spinicornis	1,148	1.34	11
Acaeroplastes melanurus	5	0.01	41
Porcellionides cingendus	1,083	1.26	12
Porcellionides pruinosus	690	0.80	15
Trachelipus rathkii	339	0.40	20
Total number woodlice records	85,926	100 %	

* species aggregates

Table 3. Individuals who have submitted over 1,000 records up to December 2007.

Recorder	No. records	Recorder (cont.)	No. records
Jon Daws	6,833	Roy Anderson	2,360
Paul Lee	3,878	Douglas Richardson	2,324
Steve Gregory	3,810	John Harper	2,083
Paul Harding	3,701	Martin Cawley	2,001
David Bolton	3,636	Peter Harvey	1,537
Arthur Chater	2,744	Paul Richards	1,494
Adrian Rundle	2,713	Steve Hopkin	1,378
Glyn & Dawn Collis	2,517	Mark Telfer	1,319
Keith Alexander	2,448	Eric Philp	1,301
Declan Doogue	2,442	David Scott-Langley	1,282

Check-list, classification and nomenclature

British and Irish check-list

The check-list of waterlice occurring in Britain and Ireland given in Gledhill, Sutcliffe and Williams (1993) includes four species. Since then no additional species have been added. Four additional species of Asellidae, described from Britain by W.E. Collinge in the 1940s, were reviewed by Moon (1953) and Harding and Moon (1976) and they are considered to be synonymous with previously described species of Isopoda.

The check-list of woodlice given in Harding and Sutton (1985) was revised and updated by Hopkin (1991a) to include thirty-seven species believed to be native or naturalised in Britain and Ireland. This included *Haplophthalmus montivagus*, *Metatrichoniscoides leydigii* and *Trichoniscoides helveticus* that were recorded for the first time. In addition to these species Oliver and Meechan (1993) also included within their checklist *Acaeroplastes melanurus*, which was believed to be extinct, and ten alien species confined to glasshouses.

The current check-list presented below includes four species of Asellidae and forty species of Oniscidea that are native or naturalised. The Oniscidea include *A. melanurus* (which has been rediscovered in Ireland), an as yet unidentified species of *Chaetophiloscia* (Jones & Pratley, 1987) and the upgrading of the two forms of the ubiquitous *Trichoniscus pusillus* aggregate into distinct species (Schmalfuss, 2004). Twelve glasshouse aliens are also included, including *Styloniscus mauritiensis*, which had been inadvertently omitted from previous check-lists (Collis & Harding, 2007) and *Venezillo parvus*, which was first collected in 2005 (Gregory, 2009).

Nomenclature of waterlice and woodlice follows the current *World List of Marine, Freshwater and Terrestrial Isopod Crustaceans* (Schotte, Boyko, Bruce, Markham, Poore, Taiti & Wilson, 2008 onwards), which for woodlice follows the *World Catalog of Terrestrial Isopods* (Schmalfuss, 2004). Gregory (2006) gives a brief overview of the nomenclature changes relevant to the British and

Irish oniscid check-list. Future updates on both check-lists will be placed on the BMIG website (www.bmig.org.uk).

As a rule genera within families, and species within genera, have been listed alphabetically, since in many cases phylogenetic relationships remain poorly understood. Valid species names are shown in bold type. The most recent synonyms, which include a number of minor spelling mistakes perpetuated by earlier British workers, are listed beneath the current name. Those marked with an * are alien species only found inside glasshouses.

Systematic check-list

CLASS CRUSTACEA
ORDER ISOPODA
      Sub-order ASELLOTA – Aquatic Waterlice
         Family Asellidae
            Asellus aquaticus (Linnaeus, 1758)
            Proasellus cavaticus (Leydig, 1871)
               Asellus cavaticus Leydig, 1871
               Asellus cavaticus Schiödte, 1871
            Proasellus meridianus (Racovitza, 1919)
               Asellus meridianus Racovitza, 1919
            Caecidotea communis (Say, 1818)
               Asellus communis (Say, 1818)

      Sub-order ONISCIDEA – Terrestrial Woodlice
         Section Diplocheta
         Family Ligiidae
            Ligia oceanica (Linnaeus, 1767)
            Ligidium hypnorum (Cuvier, 1792)
         Section Synocheta
         Family Trichoniscidae
            Androniscus dentiger Verhoeff, 1908
            Buddelundiella cataractae Verhoeff, 1930
            Haplophthalmus danicus Budde-Lund, 1880
            Haplophthalmus mengii (Zaddach, 1844)
               Haplophthalmus mengei (Zaddach, 1844)
            Haplophthalmus montivagus Verhoeff, 1941
            Metatrichoniscoides celticus Oliver & Trew, 1981
            Metatrichoniscoides leydigii (Weber, 1880)
               Metatrichoniscoides leydigi (Weber, 1880)
            * **Miktoniscus linearis** (Patience, 1908)
            Miktoniscus patiencei Vandel, 1946
               Metatrichoniscoides leydigii (Weber, 1880)
            Oritoniscus flavus (Budde-Lund, 1906)
            Trichoniscoides albidus (Budde-Lund, 1880)
            Trichoniscoides helveticus (Carl, 1908)
               Trichoniscoides helveticus Carl, 1908
            Trichoniscoides saeroeensis Lohmander, 1923
            Trichoniscoides sarsi Patience, 1908

Trichoniscus provisorius Racovitza, 1908
>> _Trichoniscus pusillus_ form _provisorius_ Racovitza, 1908

Trichoniscus pusillus Brandt, 1833
>> _Trichoniscus pusillus_ form _pusillus_ Brandt, 1833

Trichoniscus pygmaeus Sars, 1898

Family Styloniscidae
>> * **_Cordioniscus stebbingi_** (Patience, 1907)
>> * **_Styloniscus mauritiensis_** (Barnard 1936)
>> * **_Styloniscus spinosus_** (Patience, 1907)
>>> _Cordioniscus spinosus_ (Patience, 1907)

Section Crinocheta

Family Halophilosciidae
>> **_Halophiloscia couchii_** (Kinahan, 1858)
>>> _Halophiloscia couchi_ (Kinahan, 1858)
>> **_Stenophiloscia glarearum_** Verhoeff, 1908
>>> _Stenophiloscia zosterae_ Verhoeff, 1928

Family Philosciidae
>> * **_Burmoniscus meeusei_** (Holthuis, 1947)
>>> _Chaetophiloscia meeusei_ Holthuis, 1947
>> **_Chaetophiloscia_ sp.** (females only)
>> **_Philoscia muscorum_** (Scopoli, 1763)
>> * **_"Setaphora" patiencei_** (Bagnall, 1908)
>>> _Chaetophiloscia patiencei_ (Bagnall, 1908)

Family Platyarthridae
>> **_Platyarthrus hoffmannseggii_** Brandt, 1833
>>> _Platyarthrus hoffmannseggi_ Brandt, 1833
>> * **_Trichorhina tomentosa_** (Budde-Lund, 1893)

Family Oniscidae
>> **_Oniscus asellus_ ssp. _asellus_** Linnaeus, 1758
>> **_Oniscus asellus_ ssp. _occidentalis_** Bilton, 1994

Family Armadillidiidae
>> **_Armadillidium album_** Dollfus, 1887
>> **_Armadillidium depressum_** Brandt, 1833
>> **_Armadillidium nasatum_** Budde-Lund, 1885
>>> _Armadillidium speyeri_ Jackson, 1923
>> **_Armadillidium pictum_** Brandt, 1833
>> **_Armadillidium pulchellum_** (Zencker, 1798)
>> **_Armadillidium vulgare_** (Latreille, 1804)
>> **_Eluma caelatum_** (Miers, 1877)
>>> _Eluma purpurascens_ Budde-Lund, 1885

Family Armadillidae
>> * **_Reductoniscus costulatus_** Kesselyák, 1930
>> * **_Venezillo parvus_** (Budde-Lund, 1885)

Family Cylisticidae
>> **_Cylisticus convexus_** (De Geer, 1778)

Family Porcellionidae
>> * **_Agabiformius lentus_** (Budde-Lund, 1885)
>> **_Porcellio dilatatus_** Brandt, 1833
>> **_Porcellio laevis_** Latreille, 1804

Porcellio scaber Latreille, 1804
Porcellio spinicornis Say, 1818
    Porcellio pictus Brandt & Ratzeburg, 1833
Acaeroplastes melanurus (Budde-Lund, 1885)
    Metoponorthus melanurus Budde-Lund, 1885
Porcellionides cingendus (Kinahan, 1857)
    Metoponorthus cingendus (Kinahan, 1857)
Porcellionides pruinosus (Brandt, 1833)
    Metoponorthus pruinosus (Brandt, 1833)
Family Trachelipodidae
    * ***Nagurus cristatus*** (Dollfus, 1889)
    * ***Nagurus nanus*** (Budde-Lund, 1908)
    Trachelipus rathkii (Brandt, 1833)
        Trachelipus rathkei (Brandt, 1833)

Distribution maps and species accounts

Introduction to species maps and accounts

The following section provides distribution maps and species accounts of all native and naturalised species occurring in Britain and Ireland. The maps are based on records submitted to the Non-marine Isopod Recording Scheme from its onset until December 2007. The scheme has received 69,633 records for waterlice and 85,950 records for woodlice. These are plotted at a resolution of 10km using the British and Irish National Grids. Alien species, confined to glasshouses, are not mapped, but a brief account for each species is included at the end of this section.

Phenology

Although most records are made during summer months, woodlice and waterlice are found all year around. This summer peak simply reflects the traditional behaviour of field workers who tend to be most active during warm weather, even though this is perhaps the most difficult time to sample for many species. As a rule spring or autumn, when ground conditions are damp, are good times to sample for woodlice. A few, such as small soil-dwelling trichoniscids, are more readily found in winter, when they seem to be active nearer the surface and, therefore, more easily sampled. No attempt has been made to analyse the distribution of species data throughout the year.

Coverage maps

The overall coverage maps (Maps 1 and 8) show the 10km squares from which at least one record of a waterlouse or woodlouse has been received. In all probability, the absence of records from a square is due to an absence of recording rather than an absence of woodlice. The recording effort for waterlice has not been uniform (Map 1). Although there has been good coverage across much of England, records become very patchy in south-western areas and Wales and very little is known of the asellid fauna of the Republic of Ireland and Scotland. For woodlice geographic coverage across Britain and Ireland is good (Map 8). The recording scheme has received oniscid records from 3,450 10km squares in Britain and Ireland. This equates to 89% of the 3,870 10km squares that contain some land across Britain and Ireland (including the Channel Islands).

Frequency of records

The frequency of records maps (Map 2 and 9) indicate the number of records made within each 10km square. It is apparent that recording effort for asellids (Map 2) has been concentrated on the English Midlands, and to a lesser extent Essex, but is very patchy elsewhere. Map 5 gives an indication of the patchy nature of the recording effort for woodlice within each 10km square. Locally well-surveyed areas, such as Kent, Essex, Oxfordshire, Leicestershire, Derbyshire, Glamorgan and Monmouthshire, are apparent, whereas much of Scotland, particularly in the uplands, is poorly represented. Although coverage is rather uniform, Ireland remains relatively under-surveyed, particularly in recent times (as many woodlice records were made before 1983).

Species richness

The species richness maps (Map 3 and 10) indicate the number of species recorded from each 10km square. Due to the small number of asellid species involved, it is difficult to interpret the map (Map 3), although it clearly highlights the absence of recording effort in many areas. The general impression for woodlice (Map 10) is that south-eastern areas of both Ireland and Britain are more species-rich, whilst species diversity progressively declines through northern England and into Scotland. This is mainly due to warmer temperatures in the south, but also reflects the occurrence of calcareous strata in south-eastern areas of England and, to a lesser extent, Ireland. However, this trend is rather obscured by the patchy nature of the recording effort and the map, in common with the frequency of records map (Map 9), effectively highlights locally well-surveyed areas across England and Wales.

Species distribution maps

The distribution map for each species shows the recorded occurrence in Britain and Ireland plotted at 10km square resolution. The species maps are shown with records in three date classes. The choice of date classes reflects two pivotal events in the recording of isopods; the inception of the Isopod Recording Scheme in 1968 and the cut-off date (1982) for records published in *Woodlice in Britain and Ireland* (Harding & Sutton, 1985). Only the most recent record for a given 10km square is shown.

- records made from 1983 to 2007
- o records made between 1968 and 1982
- + records made before 1968

The maps are not a definitive statement of the distribution of the waterlice and woodlice of Britain and Ireland, but a presentation of our current knowledge. Distribution patterns are not static and factors, such as chance introductions and long-term climatic change, will continue to alter the observed ranges of many species. The presence of a record on a map does not necessarily indicate a viable population at that locality. The record could be the result of a single observation, perhaps of a specimen accidentally transported with garden plants, rubbish or building materials. The species, or even the habitat from which it was collected, may no longer occur at the locality.

The presence of an old (pre-1983) record may indicate that the species has been lost from a locality some time ago. This is more likely to be the case with less common species or those near the edge of their range rather than for the widespread eurytopic species. However, in many cases it is simply that the area has not been surveyed more recently. A good example is the large proportion of old records for common species such as *Oniscus asellus*, *Porcellio scaber* and *Ligia oceanica* across much of Scotland and Ireland. Whilst this gives the illusion of a decline of these common species, it is actually attributable to the lack of recording in these areas since the publication of Harding & Sutton (1985). Similarly, the occurrence of blank spaces on the maps may signify that either a given species does not occur at that particular locality or, more typically, that no-one has ever looked. The latter is true for waterlice across Ireland and Scotland. With further recording our understanding of the distribution and habitat preferences of our fauna will continue to improve.

Species accounts

Each species map is accompanied by an account of the species. The species accounts for waterlice have been updated from those in Moon and Harding (1981), and have drawn on Harding (1989), Proudlove, Wood, Harding, Horne, Gledhill and Knight (2003) and Harding and Collis (2006).

Following the publication of *Woodlice in Britain and Ireland* (Harding & Sutton, 1985) the hierarchical habitat classification was omitted from the reverse of the revised (RA 51) non-marine isopod recording card. Although recorders were encouraged to hand annotate relevant habitat information onto the record cards, the majority of records subsequently received by the recording scheme did not included associated habitat information.

Thus, the woodlice accounts have drawn on those in Harding and Sutton (1985), but have been updated using published information available on species distribution, species biology and field techniques. They have been compiled using as many sources of information as possible. In addition to formally published papers, the vast wealth of information hidden away in unpublished sources, such as the Newsletter of the British Isopod Study Group, has been incorporated. The draft species accounts have been made available to active recorders and in light of their comments many additions and corrections have been incorporated.

Each species account begins with a brief description of the species mapped, highlighting differences with superficially similar species. An outline of the distribution of each species in Britain and Ireland is also given. Characteristic habitat preferences and typical microsites for each species (where known) are described. Additional information, such as seasonality and associated species, is also given where this is helpful. Finally each account finishes with an outline of the known European and global distribution of that species.

Native and naturalised waterlice (Asellota: Asellidae)

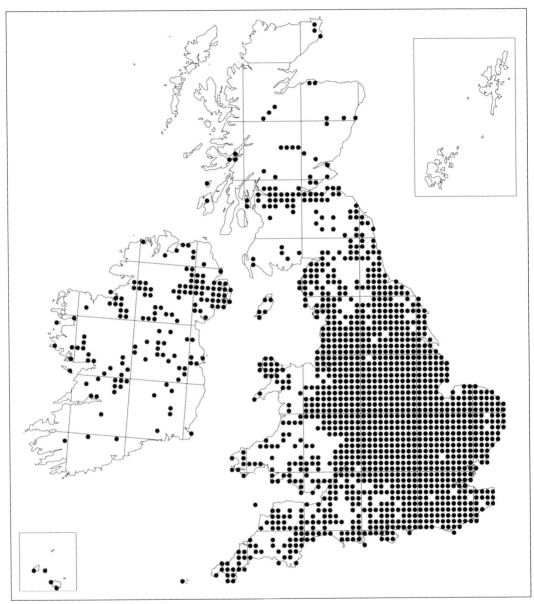

Map 1. Asellidae, overall coverage map.

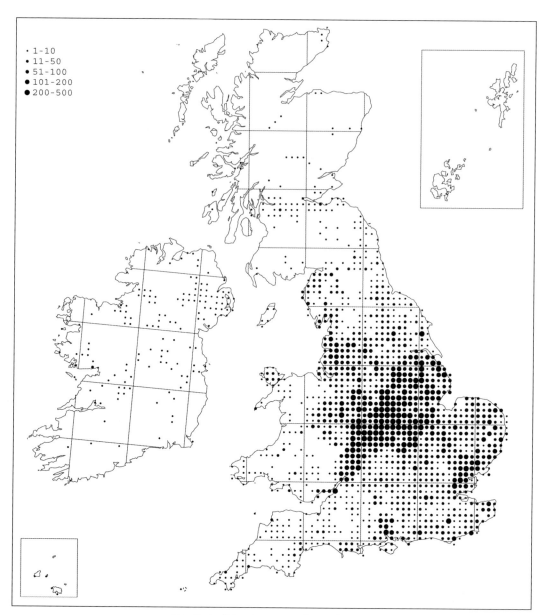

Map 2. Asellidae, frequency of records map.

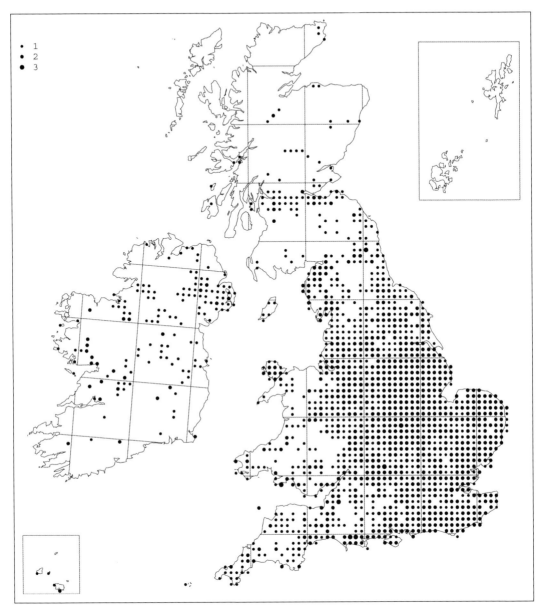

Map 3. Asellidae, species richness maps.

Asellus aquaticus (Linnaeus, 1758) and *Proasellus meridianus* (Racovitza, 1919) [*Asellus meridianus* Racovitza, 1919]

These similarly sized species, with apparently similar ecological requirements and distributions, are an important component of many freshwater ecosystems. The two species are described together, because of these similarities, but differences are highlighted.

Distinctive features: Sexually mature specimens of *Asellus aquaticus* and *Proasellus meridianus* vary between 5 and 20mm in length. Both species are mottled brown-grey in colour, but they can often be distinguished by the pattern of pigmentation of the head, even in the field. *P. meridianus* usually has a continuous pale band at the base of the head, whereas *A. aquaticus* has two pale patches.

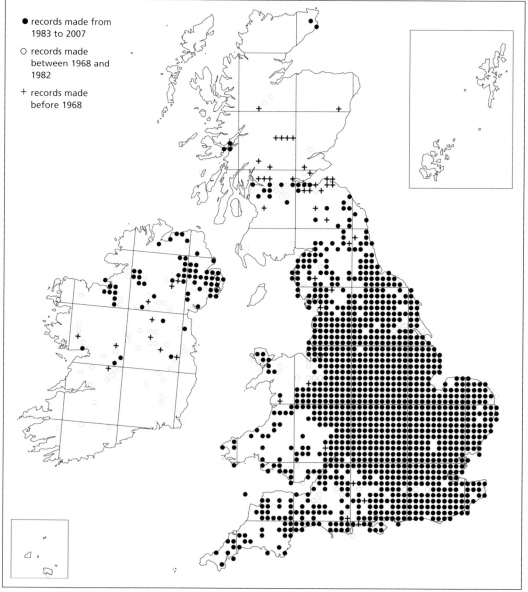

● records made from 1983 to 2007

○ records made between 1968 and 1982

+ records made before 1968

Map 4. *Asellus aquaticus*.

However, this character is variable and the pleopods or, in males, the 1st pereopods, must be checked to confirm identifications.

Distribution: A. *aquaticus* is clearly more widespread geographically than *P. meridianus*. Studies by Pond Conservation in southern England suggest that A. *aquaticus* has been found at least twice as frequently as *P. meridianus* (Jeremy Biggs, personal communication). However, on exposed western coasts and on islands *P. meridianus* is typically the dominant species.

The apparent northern and western geographical limits and scarcity of both species shown on the maps is largely a result of the absence of data from Scotland, Wales and much of Ireland. Climatic conditions are unlikely to be a limiting factor in Britain because A. *aquaticus* has been recorded in northern Norway (Økland & Økland, 1987) and inhabits lakes that are covered by ice for up

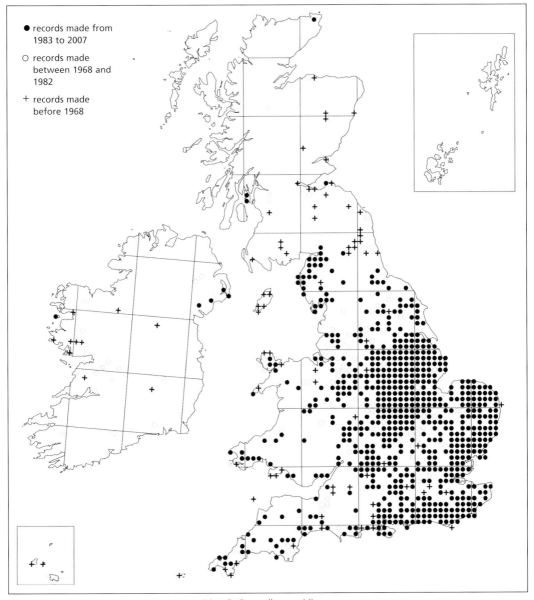

Map 5. *Proasellus meridianus.*

Figure 4. *Asellus aquaticus.* © Dick Jones.

to 8 months each year (Økland, 1980). The effects of acid deposition on upland and northern freshwaters have been suggested as a limiting factor in the spread of *A. aquaticus* in such areas (Økland, 1980).

Habitat: *A. aquaticus* appears to be capable of surviving in a greater range of habitats than *P. meridianus.* It can occur in a wide variety of water-bodies, including small urban garden ponds with artificial liners, field ponds, fenland dykes, pools in lowland marshes, canals, rivers, reservoirs, and in natural lakes as large as Lough Neagh and as deep as Windermere. Several studies have demonstrated the ability of *A. aquaticus* to survive in and adapt to polluted sites (e.g. Aston & Milner, 1980; Murphy & Learner, 1982; Maltby, 1991). *A. aquaticus* is probably more tolerant of organically polluted waters, high salinities, low pH and high metal concentrations than *P. meridianus* (Gledhill, Sutcliffe and Williams (1993). Indeed, *A. aquaticus* is often tolerant of conditions which exclude other crustaceans, for example infesting water-distribution systems and tolerating high levels of sewage contamination. *P. meridianus* appears to be less tolerant of pollution and to favour clean waters, such as those that are spring-fed. Jeremy Biggs (personal communication) observed that he has never encountered *P. meridianus* in a garden pond, but also commented that garden ponds are notoriously under-recorded.

Microsites: *A. aquaticus* and *P. meridianus* can occur among the foliage of submerged and emergent water plants, under stones and submerged bits of dead wood, among the roots of riparian trees such as alders and willows, and on the stonework of bridges. In suitable water bodies they occur almost anywhere that provides shelter from strong currents and a potential source of food among decaying plant material or on submerged surfaces.

Other notes: The distribution of these two species in Britain has been the subject of considerable interest for over 70 years. Philip Moon began his ecological studies in the English Lake District in the 1930s. In the 1950s and '60s, he published a series of papers on changes in the distribution of both species in the Lake District and adjoining areas. In the late 1950s, Bill Williams, for his PhD, also looked at the Lake District populations of both species, extending also to laboratory studies, surveys in other areas, and the collation of records that were the basis of early, county-based distribution maps (Hynes, Macan & Williams, 1960).

Both Moon and Williams examined the possible causes for the apparent decline of *P. meridianus* and increase of *A. aquaticus* in Britain. Williams (1962) regarded *P. meridianus* as the first to re-colonise Britain and Ireland after the end of the last glaciation. *A. aquaticus* was considered to have followed, to have competed with and, in places, to have replaced *P. meridianus*. The dominance of *P. meridianus* at sites on exposed western coasts and on islands was taken to indicate the retreat of this species in the face of competition from *A. aquaticus*. Later work by Williams (1979) provided only qualified support for interspecific competition and replacement of *P. meridianus* by *A. aquaticus*. Moon and Harding (1981), drawing on the work of Sutcliffe (1974) and others, suggested that water chemistry, especially the sodium content of waters, was probably more important than competition. Considering its tolerance to pollution, the success and apparent spread of *A. aquaticus* is probably just another example of the degradation of the semi-natural environment being caused by humans.

A study of weed-control measures of canals in Ireland, using mechanical cutting and application of a herbicide (dichlobenil), showed only short-term effects on numbers of *A. aquaticus* (Monahan & Caffrey, 1996). A laboratory study of the effects of the application of an insecticide (chlorpyrifos) on detritivorous macroinvertebrates in nutrient enriched water showed that established populations of both *A. aquaticus* and *P. meridianus* collapsed (Cuppen, Glystra, van Beusekom, Budde & Brock, 1995).

The spread of either species, but particularly *A. aquaticus*, is also probably contributed to by inadvertent transfer by humans, for example with water plants for garden ponds and with fish stocks for angling. Large scale flooding events on floodplains, such as those of the Severn and Humber in 2007, must also contribute to the potential spread of *A. aquaticus*. Other environmental changes could affect populations of both species adversely.

Worldwide distribution: *A. aquaticus* is a very widespread European species, occurring from Scandinavia east to Russia and as far south as Italy, Greece and Turkey (Gruner, 1965). *P. meridianus* has a more restricted distribution centred on north-western Europe, occurring within France, Belgium, Netherlands and Germany (Gruner, 1965).

Proasellus cavaticus (Leydig, 1871)
Asellus cavaticus Leydig, 1871
Asellus cavaticus Schiödte, 1871

Distinctive features: P. *cavaticus* is eyeless and devoid of pigment. However, not all waterlice occurring underground and with little or no colour should be assumed to be *P. cavaticus*, because almost colourless specimens of A. *aquaticus* and *P. meridianus* have occurred in caves. Sexually mature specimens normally range from 3 to 8mm.

Distribution: This species was first recorded in Britain in 1925, from a well at Ringwood, Hampshire. There was an active phase of biological recording in caves and mines, mainly in the

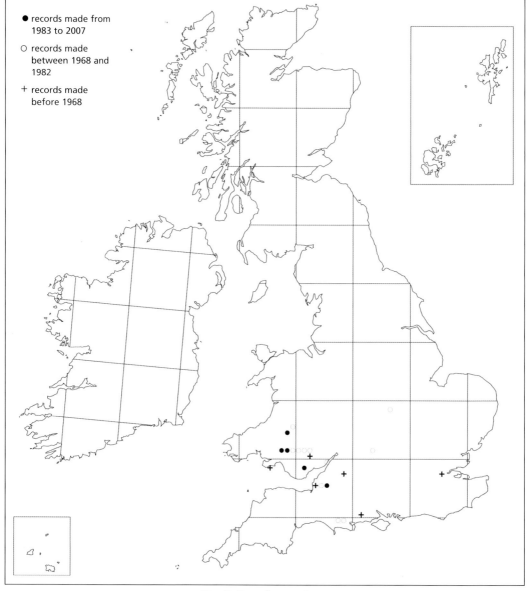

● records made from
  1983 to 2007

○ records made
  between 1968 and
  1982

+ records made
  before 1968

Map 6. *Proasellus cavaticus*.

1950s and 1960s, when most of the records of *P. cavaticus* in Britain were made. The majority are from carboniferous limestone cave systems in the Mendip Hills and South Wales. Despite surveys of similar cave systems in Derbyshire and North Yorkshire, *P. cavaticus* has not been recorded in these northern caves. The most northerly records are in mid-Wales on the Hirnant River (Ormerod & Walters, 1984) and at Nant Esgair Gars.

Habitat: This species is a specialist stygobite and has been recorded from underground streams, pools and wet surfaces in limestone caves and mines, and from wells and bore-holes. When found above ground (such as the records at Hirnant River and at Nant Esgair Gars) it is always where underground waters issue on the surface, such as springs and the water sources of watercress beds. These are probably vagrant specimens washed to the surface.

Microsites: Observations in the Ogof Ffynnon Ddu cave system in south Wales have been summarised by Jefferson, Chapman, Carter and Proudlove (2004). Here *P. cavaticus* is largely sedentary and usually seen in small numbers. Typically one or two individuals may be encountered in water films covering flowstones, within small seepages or in shallow pools. Specimens may also occur beneath stones. In these situations there is adequate food supply, such as organic debris and bacterial films, but little risk of being washed away by floodwater and fewer predators, such as the Amphipod *Niphargus fontanus*. Above ground records are typically from among gravel on stream or river beds, near springs or seepages. At Nant Esgair Gars specimens were collected at 50cm depth in gravel.

Other notes: Some populations exhibit size differences, with a particularly small morph (to 4mm) occurring in the vadose zone of caves in the Mendip Hills, Somerset, although a much larger morph occurs in the phreatic zone in the same area (Proudlove, Wood, Harding, Horne, Gledhill & Knight, 2003). Continental authors have described several subterranean *Proasellus* taxa as distinct species, most of which occur in biogeographically isolated cave systems. If size variation in British populations is genetically linked, then the distinct Mendip populations are likely to be separate taxa (Proudlove *et al*, 2003). Additionally, some of the cave systems in south Wales (e.g. Ogof Ffynnon Ddu) are thought to be over 1 million years old and if relict populations of *P. cavaticus* had survived within pre-existing caves beneath the glacial ice sheets then these too may represent another distinct taxon (Proudlove *et al*, 2003).

Moon & Harding (1981) noted the loss of contact with underground waters that resulted from the abandonment of domestic wells in favour of piped water supplies. Since then, ground-waters have been increasingly exploited to supply water for domestic and commercial purposes and pollution of ground-waters by biocides and fertilizers may have become a problem in some areas. Concern about human impacts on subterranean Crustacea in Belgium was expressed by Fiers and Wouters (1985), especially the effects of eutrophication on ground-waters and general disturbance in cave systems open to the public. Proudlove *et al* (2003) briefly considered the issues relating to the conservation of subterranean aquatic crustaceans in Britain.

Worldwide distribution: *P. cavaticus* occurs widely in continental Europe, from the Netherlands east to the Hartz Mountains and the Danube in Austria, and from the Alps in Germany. It has not been recorded from Ireland.

Caecidotea communis (Say, 1818)
Asellus communis (Say, 1818)

Distinctive features: In the field, *Caecidotea communis* is superficially similar to *A. aquaticus* and *P. meridianus*. It is generally rather larger than both these species and has a looser, 'floppy' appearance (Harding & Collis, 2006). Describing the neotype from the USA, Williams (1970) gave a length of 11mm, but recorded males ranging from 4 to 18mm.

Distribution: This is a North American species first discovered in Britain in the early 1960s, at Bolam Lake, Northumberland (Sutcliffe, 1972; Williams, 1972). Bolam Lake, and a short length of a small outfall stream from the lake, remains the only known locality in Britain for *C. communis* (Harding & Collis, 2006).

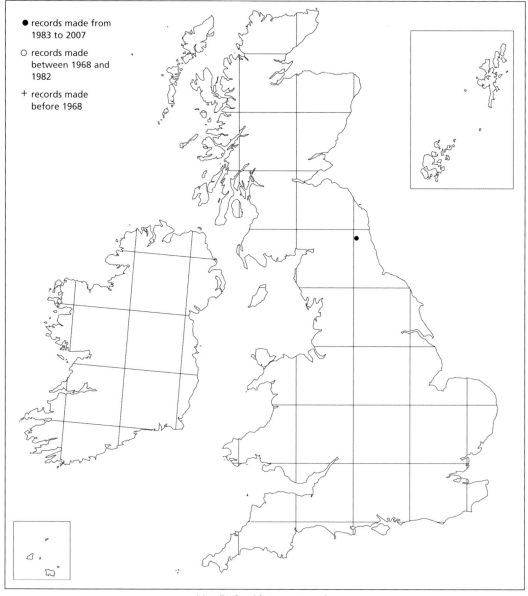

● records made from
1983 to 2007

○ records made
between 1968 and
1982

+ records made
before 1968

Map 7. *Caecidotea communis*

Habitat: Bolam Lake was created in the early 19th Century by deepening and damming a boggy area fed by small streams and springs. The 10 hectare lake is the central feature of the Bolam Lake Country Park. The habitat in which Harding and Collis (2006) collected *C. communis* was similar to that described by Sutcliffe (1972): in generally clear water with a substrate of mud or fine gravel and large quantities of decaying leaves from trees and shrubs. Water plants, such as Canadian waterweed *Elodea canadensis*, another introduced North American species, were present but not abundant. Magnin and Leconte (1971, 1973) described the life cycle of *C. communis* in North America, and Kaushik and Hynes (1971) stated that it feeds avidly on the decaying leaves of a range of broadleaved trees.

Microsites: Harding and Collis (2006) give a full account of the occurrence of *C. communis* in Bolam Lake, based on their survey in April 2005. Specimens of *C. communis* were collected at several points around the lakeshore, most plentifully in slightly deeper water (about 1 metre) with a fine gravel substrate, around the fishing pier in Low House Wood.

Associated species: Harding and Collis (2006) noted that all specimens of Asellidae examined at Bolam Lake proved to be *C. communis*. Other fauna observed in the samples included leeches, molluscs and amphipods, including *Crangonyx pseudogracilis*, a North American species that is now widespread in Britain and parts of Ireland. Sutcliffe (1972) had also noted that *A. aquaticus* and *P. meridianus* were not recorded at Bolam Lake in the 1960s.

Worldwide distribution: *C. communis* is indigenous and widespread in North America and occurs over a wide range of latitudes. Williams (1970) reports records from Canada (Nova Scotia, Ontario) and the USA (from West Virginia to Maine in the east, plus Colorado and Washington State). There it occurs in a wide variety of inland waters, both still and flowing, but it has not been recorded in numerous collections from the Great Lakes. It is also reported to be widespread in central Mexico (Cole & Minkley, 1968).

Native and naturalised woodlice (Oniscidea)

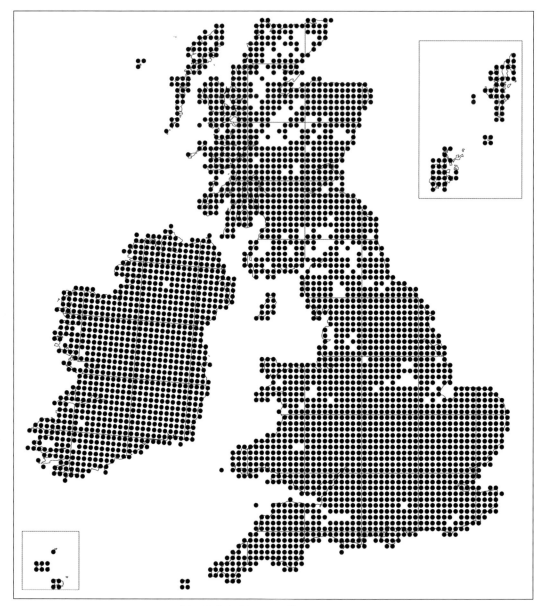

Map 8. Oniscidea, overall coverage map.

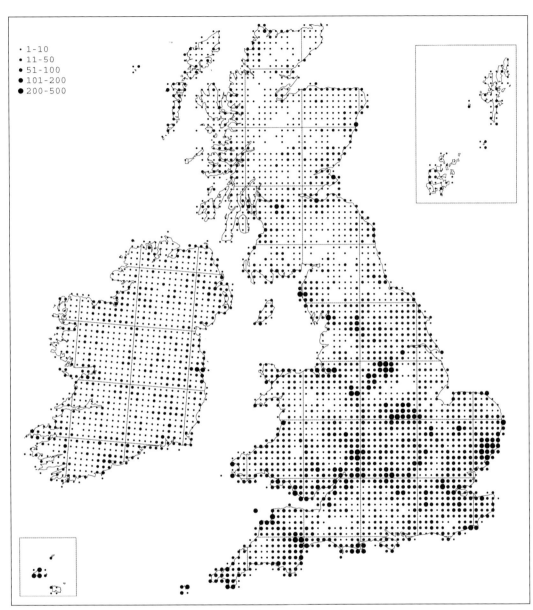

Map 9. Oniscidea, frequency of records map.

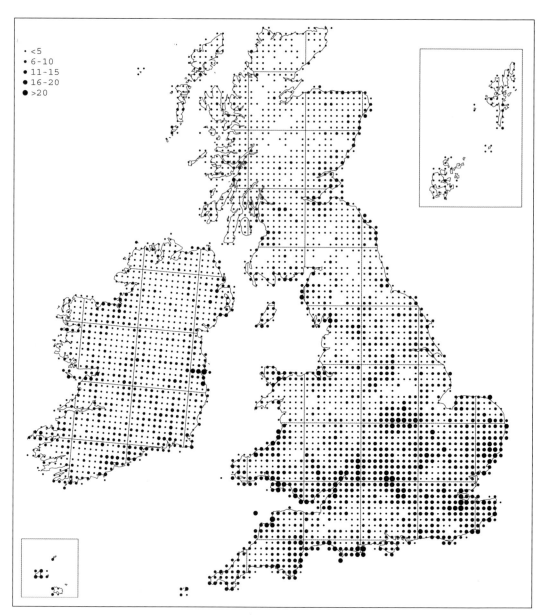

Map 10. Oniscidea, species richness map.

Ligia oceanica (Linnaeus, 1767)

Distinctive features: This very large species (up to 30mm in length) is remarkably fast and agile. The antennal flagellum is composed of numerous bead-like segments. In the field immatures may be confused with *Halophiloscia couchii*, which is distinguished by having three elongated flagella segments.

Distribution: Around the entire coastline of Britain and Ireland *L. oceanica* has proved to be common wherever rocky shoreline occurs, including virtually every offshore island visited. It also penetrates the tidal reaches of many estuaries. The predominance of records in north-western Scotland simply reflects the convoluted nature of the fjordic coastline there. In areas where the coastline is 'soft', such as Suffolk (Lee, 1993), the species becomes much more scarce. The inland Yorkshire record, from a Bradford warehouse, refers to an imported specimen (Harding, 1976a).

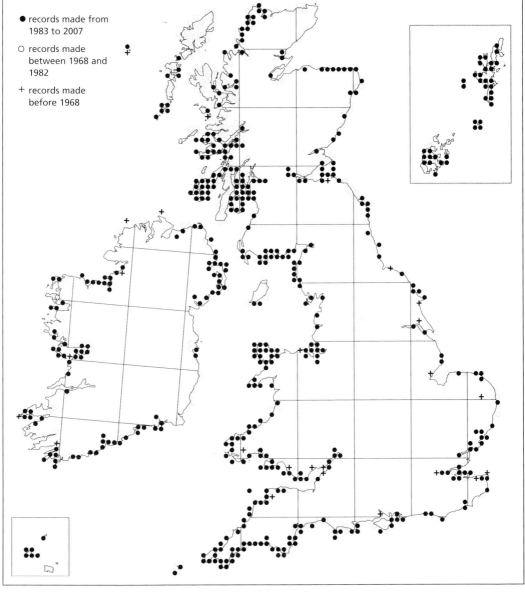

Map 11. *Ligia oceanica*.

Figure 5. *Ligia oceanica*. © Dick Jones

Habitat: At the base of unvegetated rocky cliffs this can be the most abundant woodlouse, often to the exclusion of other species. Where there is shelter from extremes of storms and tides it can also be found on boulder beaches, shingle beaches or erosion banks, but it generally avoids sandy beaches and salt marsh. On 'soft' coastlines it is invariably associated with solid man-made structures, such as harbour walls and jetties. It is usually found in the littoral zone, often around the high tide mark, where the substrate is devoid of vegetation. It is tolerant of immersion in seawater and can occur much lower down the beach than other woodlice (which tend to inhabit the supralittoral zone and above). On occasions it can be found a considerable distance up unvegetated cliffs, particularly where exposed conditions produce plenty of salt spray.

Microsites: It is most easily found by turning boulders and large stones that have accumulated at the base of sea cliffs. However, on 'solid' features, such as sea cliffs or harbour walls, it occupies inaccessible crevices during the day and can be extremely difficult to find. In this situation the species may be readily found at night (Harding & Sutton, 1985) when it is active on the surface. Rarely, it has been found amongst shingle and strandline debris.

Associated species: Although it frequently occurs to the exclusion of other woodlice, on southern and western coastlines it is typically associated with *Halophiloscia couchii*. Where it occurs in the sparsely vegetated upper parts of the shore it may be found with species such as *Porcellio scaber* or *Armadillidium vulgare*.

Worldwide distribution: This widespread species occurs along the entire coast of western Europe from southern Iceland and the western Baltic Sea in the north to the Straits of Gibraltar in the south (Schmalfuss, 2004; Easter, 2005).

Ligidium hypnorum (Cuvier, 1792)

Distinctive features: A medium sized fast-moving woodlouse, up to 9mm in length. Confusion is most likely with the common *Philoscia muscorum*, which is of a similar size and shape. *L. hypnorum* is much more attractively mottled and has the antennal flagellum composed of numerous bead-like segments (three in *P. muscorum*).

Distribution: The south-eastern distribution indicated in Harding and Sutton (1985) has been bolstered by a considerable quantity of records from Essex and Suffolk. East of a line from the Isle of Wight to the Wash, this can be a locally common woodlouse. It is apparent that its range extends much further south-west than previously thought, with outlying populations recently

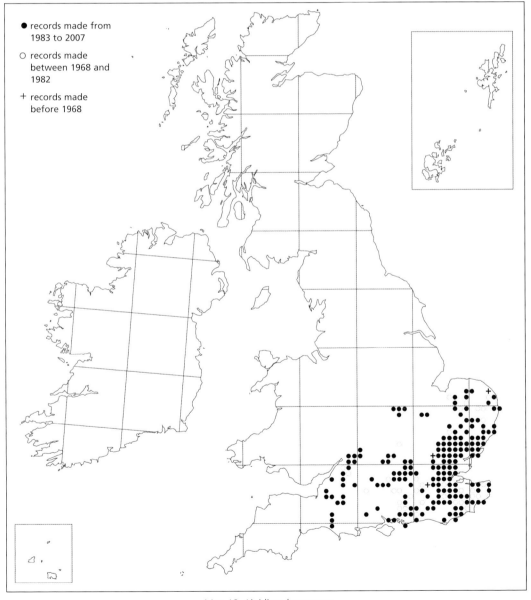

Map 12. *Ligidium hypnorum*

Figure 6. *Ligidium hypnorum.* © Theodoor Heijerman

discovered in South Somerset and West Dorset. This is undoubtedly due to increased recorder activity in these areas, rather than an actual expansion of this species' range. The north-western limit, southern Leicestershire, remains unchanged.

Habitat: This hygrophilous species thrives in waterlogged habitats and is tolerant of temporary inundation (Holdich, 1988). It is equally tolerant of lime-rich or non-calcareous conditions. It is most abundant in eastern areas, where it can be found in a wide range of habitats from wetlands to damp meadows and deciduous woodlands, often beside watercourses. Further west it becomes more restricted to semi-natural habitats, such as ancient woodlands and fen carr. At its most northerly British locality in Leicestershire it is associated with tall herb fen vegetation beside the river Welland (Daws, 1997). It is rarely found in synanthropic or coastal habitats, leading Harding and Sutton (1985) to suggest that it is a relict forest species.

In Oxfordshire, it occupies three distinct types of ancient woodland (Gregory & Campbell, 1995). In fens or on river floodplains it can be locally abundant in mature sallow *Salix* or alder *Alnus* carr. In the poorly drained woodlands of the clay vales it favours damp hollows or the edges of watercourses. In the dry, well-drained chalk woodlands of Chiltern Hills, it occurs in small numbers, but only on shady north-facing slopes.

Microsites: It is typically found amongst moss and leaf litter or within sedge and grass tussocks, but also beneath dead wood and stones. Being an active surface-dwelling species it can be numerous in pitfall trap samples.

Associated species: It is usually found with other species characteristic of wet habitats, including *Haplophthalmus danicus, Trichoniscus pusillus* agg., *Trachelipus rathkii,* and in inclement weather, *Trichoniscoides albidus.* In drier places it may be found with species such as *Philoscia muscorum.*

Worldwide distribution: The range of this extremely widespread species extends eastwards through central Europe and into western Asia (Schmalfuss, 2004).

Androniscus dentiger Verhoeff, 1908

Distinctive features: This is a small (up to 6mm) fast-moving species. It is variable in colour, but typically bright orange or salmon pink. The entire body is covered in coarse tubercles, easily seen with a hand lens. Any pigmentation is rapidly lost in alcohol, but the single black ocelli remain prominent. Immatures may be mistaken for *Trichoniscoides* spp. (which have reddish ocelli that fade in alcohol) and *Miktoniscus patiencei* (which is pure white in colour).

Distribution: Across much of lowland Britain *A. dentiger* is a widespread and common woodlouse. It has been recorded from all 40 vice-counties in Ireland (Cawley, 2001). The distribution map has considerably filled in since the publication of Harding and Sutton (1985), but the species

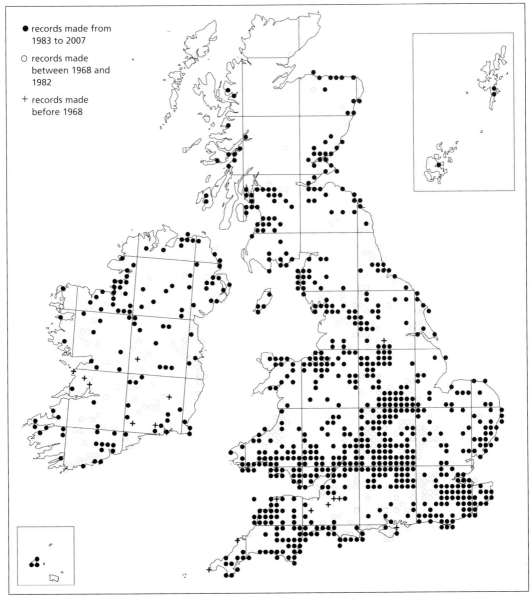

● records made from
1983 to 2007

○ records made
between 1968 and
1982

+ records made
before 1968

Map 13. *Androniscus dentiger.*

Figure 7. *Androniscus dentiger.* © Paul Richards.

remains under-recorded in many areas. It seems to be genuinely scarce in Scotland and in upland areas of northern England, suggesting it becomes restricted by altitude and latitude, but in south Wales it occurs to 585m in the Brecon Beacons (Morgan, 1994). The coastal records from Orkney and Shetland presumably reflect the amelioration of latitude by maritime influences.

Habitat: This species is equally at home in semi-natural coastal habitats and synanthropic sites. On the coast A. *dentiger* can be found amongst sparsely vegetated shingle or on boulder beaches and erosion banks above the extreme high tide mark. Away from the coast it is not often found in semi-natural habitats. However, it can be locally frequent where carboniferous limestone occurs, such as Derbyshire (Richards, 1995) and Carmarthenshire (Morgan, 1994). This species is also found in a wide array of synanthropic habitats, including churchyards, gardens (including underground drains and buried cable ducting), cellars, waste ground, disused quarries, farmyards, rubbish tips, and disused railway lines. Most of the northern and western records, such as Shetland (Daws, 1994b), Isle of Mull and Inner Hebrides (Scott-Langley, 2002), are synanthropic coastal sites. A. *dentiger* is a well-documented troglophile and is the most frequently encountered woodlouse in caves, potholes, tunnels, mines and sewers.

Microsites: Typically this species is found on the underside of large stones or amongst rubble. On the coast it may be found under strandline debris or within crevices several centimetres down where peaty soil has accumulated. On sparsely vegetated 'soft rock' sea cliffs it is often associated with seepages and other wet spots and is most easily seen after dark (Harding & Sutton, 1985). It can also be found under or within rotten dead wood, in compost heaps and in crevices of old mortared walls. It tolerates much drier places than other trichoniscids, yet conversely it is frequently found in wet places, even completely submerged at stream margins (Daws, 1994a) and deep inside caves (Chapman, 1993). Cawley (1996) records its occurrence in wet soil beside streams.

Other notes: It is able to maintain thriving populations deep underground and research in Italy has indicated that cave populations there are genetically isolated from each other and from those at the surface (Gentile & Sbordoni, 1998). Permanent cave-dwelling populations occurring in Britain and Ireland, if shown to be genetically distinct, will be of high conservation status.

Associated species: Typically, it is associated with other soil-dwelling trichoniscids, such as *Trichoniscus pusillus* agg., *Trichoniscus pygmaeus* and *Haplophthalmus* spp. In subterranean habitats it has been observed with *Trichoniscoides saeroeensis* (Moseley, 1995) and the asellid *Proasellus cavaticus* (Chapman, 1993).

Worldwide distribution: This species occurs widely across Western Europe and North Africa, often in synanthropic habitats, and has been introduced into North America (Schmalfuss, 2004). It is apparently absent from the Iberian peninsula.

Buddelundiella cataractae Verhoeff, 1930

Distinctive features: This is a very small white to pinkish-buff woodlouse reaching 3mm. Its ability to roll into a ball and the exaggerated haplophthalmoid sculpturing make *B. cataractae* readily identifiable. Confusion is most likely with the glasshouse alien *Reductoniscus costulatus*.

Distribution: This apparently very rare species was first discovered in Britain in 1981 (Oliver, 1983). Harding and Sutton (1985) listed three sites in south Wales, in Glamorgan, and a fourth at Snettisham, West Norfolk, but suggested that other coastal localities may await discovery. *B. cataractae* remains elusive and just seven additional sites have been discovered to date: Oxford city, Oxfordshire(1989); Eastbourne, Sussex (1993); Clydach Gorge, Monmouthshire (2004);

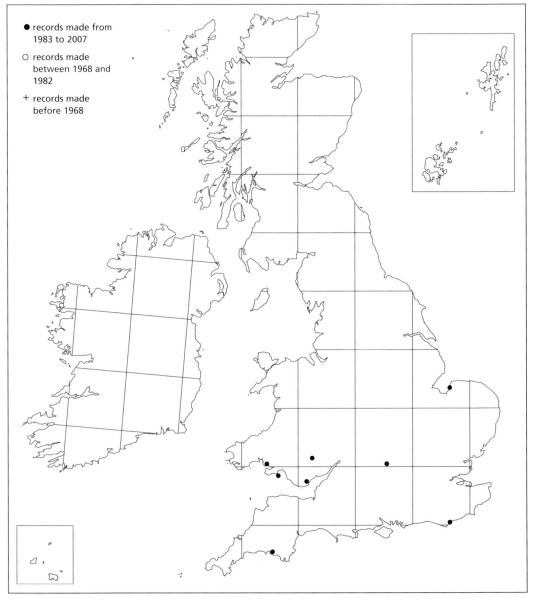

Map 14. *Buddelundiella cataractae*.

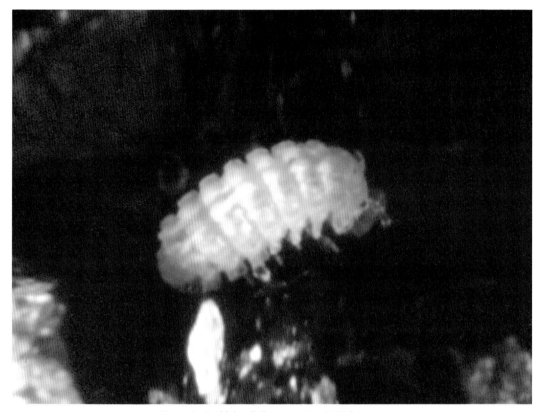

Figure 8. *Buddelundiella cataractae.* © Dick Jones

Plymouth, Devon (2005) and most recently at Mumbles Head, Glamorgan and Llanelli, Carmarthenshire in south Wales (2008). Although the majority of known sites are coastal it is clear that B. *cataractae* is capable of surviving away from maritime influences.

Habitat: Semi-natural sites include the vegetated upper parts of a shingle bank at Snettisham and, in south Wales, a boulder beach at Barry and riverside gravels in Cardiff. The site at Mumbles Head, where the species occurred amongst stones embedded in a clay bank above a shingle beach, had synanthropic influences. At Eastbourne, Sussex, it occurred on an area of sparsely vegetated waste ground beside the sea (J.P. Richards, personal communication). Clearly synanthropic sites include Oxford city, where B. *cataractae* was found at a garden centre on the site of a former plant nursery (Gregory & Campbell, 1995) and a garden at Cardiff (Oliver, 1983). The Plymouth site is a disused railway line.

Microsites: The common theme between most sites is some degree of disturbance, either synanthropic or natural, and the presence of damp, highly organic, friable soil often several centimetres below the surface layers. At Eastbourne B. *cataractae* was found under drift wood embedded in sandy soil overlying shingle (J.P. Richards, personal communication). It occurred amongst peaty debris beneath paving stones at Oxford (Gregory & Campbell, 1995). This is akin to the Cardiff garden, where it was found under stones and loose paving (Oliver, 1983). A single specimen was found at Clydach Gorge, under a large stone resting on soil in a limestone quarry (Harper, 2004b). A specimen has been collected from beneath a decaying railway sleeper at the Plymouth site (D.T. Bilton, personal communication).

Associated species: It is usually associated with other soil-dwelling trichoniscids, typically *Haplophthalmus mengii* and *Trichoniscus pygmaeus*, and on the coast *Trichoniscoides saeroeensis*. At Eastbourne it was associated with *Eluma caelatum*.

Other notes: It is clear that *B. cataractae* is very elusive and probably extremely under-recorded. Its subterranean habits, small size and buff colouration make it easily overlooked in the field, particularly when enrolled amongst soil or shingle. Specialist collecting techniques, such as sieving, have been recommended to find this species (Irwin, 1982). It probably awaits discovery at other coastal localities and synanthropic sites inland, at least in southern England and south Wales.

Although, populations at some sites appear to be semi-natural, others are clearly synanthropic. It is possible that most, if not all, of the British localities are introductions. Harper (2004b) considers the species to have been introduced to Clydach Gorge via garden rubbish. However, if not native in Britain, *B. cataractae* is certainly well established and naturalised. It has undoubtedly been introduced at the Oxford site and Gregory (2001) suggests that it may have been present, unnoticed, for many decades. This site (which also supports *Metatrichoniscoides leydigii*) has been redeveloped for housing and it will be interesting to see if *B. cataractae* survives.

Worldwide distribution: This elusive species has been recorded barely a dozen times in continental Europe, but is widely distributed from the Mediterranean north to Finland and east to Georgia (Oliver & Meechan, 1993).

Haplophthalmus danicus Budde-Lund, 1880

Distinctive features: This small white to buff species (to 4mm) has eyes composed of a single black ocellus. The longitudinal dorsal ridges, which are characteristic of the genus, are much less distinct than seen in *H. mengii* and *H. montivagus*. It also lacks the pair of prominent projections on the third pleonite seen in these two species.

Distribution: The map suggests a predominantly south-eastern distribution, but it is clear that there is much recorder bias. Experience in well-worked areas, such as Kent, North Essex, East Suffolk and Oxfordshire, indicates that this is a common species in many areas south of a line from The Wash to the Severn Estuary. North of this line there are scattered records north to South

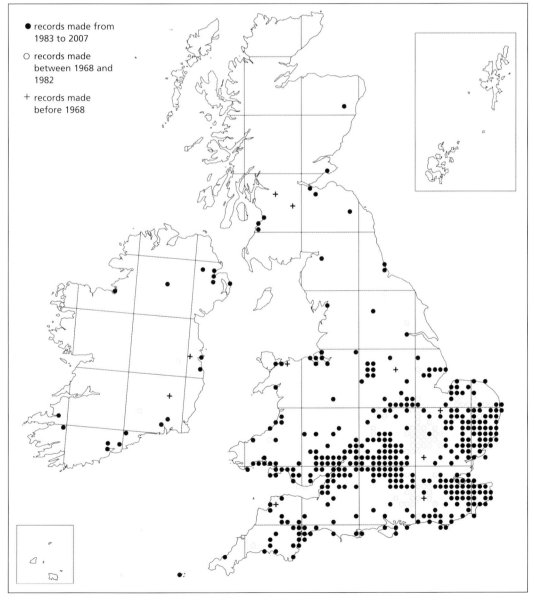

Map 15. *Haplophthalmus danicus*.

Figure 9. *Haplophthalmus danicus.* © Theodoor Heijerman

Aberdeenshire. Unlike many species, it does not become noticeably coastal towards the northern limits of its range. In Ireland there is a thin scatter of records, mainly in the east and across southern counties (Cawley, 2001). The paucity of records in well-worked Co. Sligo, in the northwest, suggests it may be genuinely scarce there (Cawley, 1996).

Habitat: This species is primarily associated with damp woodlands, both semi-natural and recent plantations, or other damp well-vegetated habitats, including reedbeds, fens, river flood plains and lakesides. It does occur in drier localities, but typically within damp hollows or besides watercourses. Synanthropic habitats, such as waste ground, farmyards, road verges, churchyards and gardens, are also utilised. *H. danicus* is often found inside glasshouses and is common inside many of the glasshouses at Kew Gardens, Surrey. In Ireland the species is primarily associated with disturbed woodland and synanthropic sites (Cawley, 2001), but is much less frequently encountered than *H. mengii*. It may have a preference for calcareous areas, but is not strongly associated with coastal habitats.

Microsites: Unlike *H. mengii*, it prefers to inhabit above-ground microsites and is characteristically found associated with damp dead wood. It can be abundant amongst peaty debris that has accumulated beneath loose bark on dead wood, beneath or within rotten logs and stumps or within compost heaps. At the most northerly British locality, Castle Fraser, South Aberdeenshire, it occurred within a well-rotted compost heap (M.B. Davidson, personal communication). *H. danicus* can also be collected from well-rotted woodland litter, wet waterside litter and damp friable humus-rich soil or found under stones where such conditions occur.

Associated species: It is typically associated with other trichoniscid woodlice, such as *Trichoniscus pusillus*, *H. mengii* and *H. montivagus*.

Other notes: At many sites it can occur at low densities and is easily overlooked due to its small size and sluggish movements. However, populations exhibit an ability to increase rapidly in numbers under suitable conditions (Harding & Sutton, 1985) and at favoured sites it can be abundant and more readily found.

Worldwide distribution: *H. danicus* is widely distributed throughout Europe and has been introduced to many other parts of the world (Schmalfuss, 2004).

Haplophthalmus mengii aggregate
Haplophthalmus mengei sensu lato

Taxonomic note: This is the *Haplophthalmus mengei* of Harding and Sutton (1985) and earlier British workers. Hopkin and Roberts (1987) have shown this to be comprised of two closely allied species, the true *H. mengii* (Zaddach) and *H. montivagus* Verhoeff. The two species can only be separated by examination of a male specimen and undifferentiated records have been lumped together as the *H. mengii* aggregate. Often this is because only female specimens have been collected, but in many cases it is because the majority of recorders do not routinely dissect male specimens. Prior to the discovery of *H. montivagus* in England specimens were rarely collected since an experienced recorder could usually differentiate "*H. mengei*" from *H. danicus* in the field.

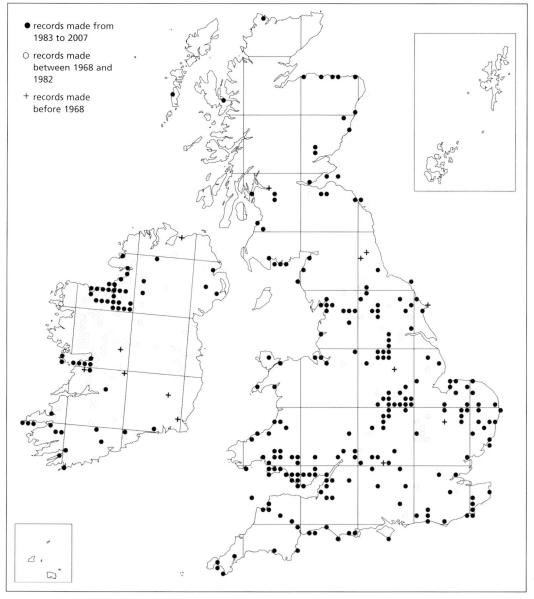

● records made from
   1983 to 2007

○ records made
   between 1968 and
   1982

+ records made
   before 1968

Map 16. *Haplophthalmus mengii* aggregate.

Distinctive features: *H. mengii* agg. is similar in general appearance to *H. danicus*. However, the longitudinal dorsal ridges are much more prominent and there is a distinct pair of dorsal projections on the third pleonite. Unfortunately, the tergal sculpturing of *H. mengii* and *H. montivagus* is identical and the two can only be separated by microscopic examination of a male specimen (hence the past confusion).

Distribution: The map shows records for *H. mengii* and/or *H. montivagus* where a male specimen has not been microscopically examined. Records for the *H. mengii* aggregate occur throughout Britain and Ireland. Although the records plotted here, and those in Harding and Sutton (1985), could refer to either species it is likely that the majority of records, and probably all of those in Ireland and Scotland, refer to the genuine *H. mengii sensu stricto*. Hopkin (1991a) suggested that *H. montivagus* and *H. mengii* might have mutually exclusive distributions in Britain. *H. mengii* does seem to become relatively scarce and predominantly coastal in south-eastern England, where it may be partially displaced by *H. montivagus* (Hopkin & Roberts, 1987). However, surveys in Oxfordshire (Gregory, 2001) and Monmouthshire (Harper, 2002) indicate that the two species can co-exist, though apparently in separate microsites.

Haplophthalmus mengii (Zaddach, 1844)
Haplophthalmus mengei (Zaddach, 1844)
Haplophthalmus mengii seg.
Haplophthalmus mengii sensu stricto

Taxonomic note: The *H. mengei* described in Harding and Sutton (1985) comprises two closely related species (Hopkin & Roberts, 1987): the true *H. mengii* (Zaddach) detailed here and *H. montivagus* Verhoeff. Records published before 1987 could refer to either species and, unless supported by a voucher specimen, have been included with the map for *H. mengii* aggregate. It is probable that the majority of records plotted in Harding and Sutton (1985) refer to the true *H. mengii* detailed here.

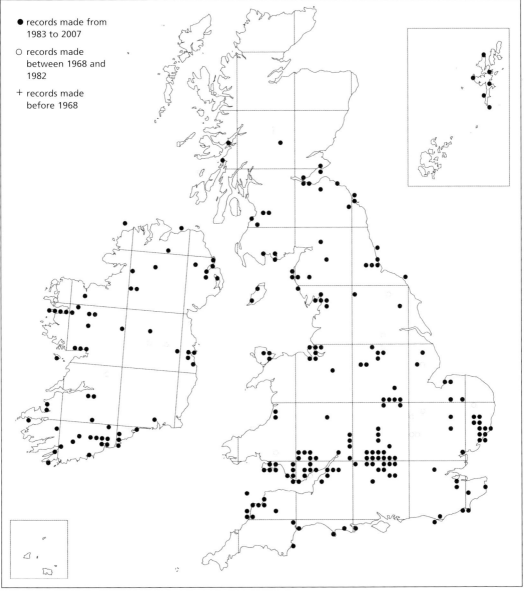

● records made from 1983 to 2007

○ records made between 1968 and 1982

+ records made before 1968

Map 17. *Haplophthalmus mengii.*

45

Figure 10. *Haplophthalmus mengii.* © Paul Richards.

Distinctive features: This dorsal sculpture is identical in *H. montivagus* and only male specimens can be identified with certainty.

Distribution: The map plots only those records verified by examination of male specimens. Records occur across the full breath of Britain, from Devon to Kent in the south, throughout England, Wales and Scotland to the Shetland Islands in the far north. The relatively few confirmed records indicate that the species is also widely distributed throughout Ireland. *H. mengii* has a north-western bias in Britain (Hopkin, 1987a) and it becomes relatively scarce in south-eastern England. The apparent southern bias, the clumped nature of records and its absence from much of Scotland and Ireland are entirely due to recorder bias. The map simply indicates areas where a few key recorders have been active since the 1980s. It should be expected to occur around the entire coastline of Britain and Ireland and to be much more widely distributed across inland counties.

Habitat: This species inhabits a wide range of semi-natural and synanthropic habitats, both inland and on the coast. Records submitted to the recording scheme indicate that the habitat preferences of *H. mengei* detailed in Harding and Sutton (1985) provide an accurate account for the true *H. mengii* described here. In coastal habitats it is fairly common on shingle, boulder beaches, erosion banks and soft cliffs and often associated with storm strandline debris. It readily takes to synanthropic sites such as farmyards and churchyards, often within compost heaps. On the Shetland Islands it was collected from synanthropic habitats (i.e. churchyard, cemetery and farmyard), but not from coastal or semi-natural habitats (Daws, 1994b).

It has a strong association with alluvial soils and in Oxfordshire it is a frequent inhabitant of riverside meadows, scrub and wet woodland along the Thames Valley (Gregory & Campbell,

1995). It also occurs widely in calcareous woodland (a habitat favoured by *H. montivagus*), especially on the Carboniferous limestones of northern England (Harding & Sutton, 1985). Gregory and Campbell (1995) report its occurrence on short-turf chalk grassland during heavy frosts. In Ireland it occurs in a similar array of semi-natural habitats, including grasslands on calcareous glacial deposits and lakeside grasslands (Doogue & Harding, 1982).

Microsites: This is a soil-dwelling species favouring calcareous soils. Although elusive, especially in dry weather, numerous specimens may be found once the exact microsite is discovered. It is typically found on the underside of deeply embedded stones and dead wood. It also occurs among rubble, friable humus-rich soil, well-rotted compost and, on the coast, within damp peaty soil that accumulates below shingle or boulders.

Associated species: It is invariably associated with other trichoniscids, such as *Haplophthalmus danicus*, *Trichoniscoides albidus*, *Trichoniscus pusillus*, *T. pygmaeus*, and on the coast, *Miktoniscus patiencei* and *Trichoniscoides saeroeensis*.

Worldwide distribution: *H. mengii* has a wide distribution throughout Europe and North Africa, but its precise range is unclear due to confusion with *H. montivagus* in many countries (Schmalfuss, 2004).

Haplophthalmus montivagus Verhoeff, 1941

Taxonomic note: H. *montivagus* was first recognized in Britain in 1987 when male specimens were found 'hiding' among reference collections of H. *mengii* (Hopkin & Roberts, 1987). Therefore, the species was unknown to Harding and Sutton (1985) and a few records (e.g. TQ81 and SP50) were inadvertently included within the map for H. *mengei* (Zaddach). Other published records for H. *mengei* (Zaddach) made prior to 1987 may also refer to H. *montivagus*.

Distinctive features: It is identical in appearance to H. *mengii* and only male specimens can be identified.

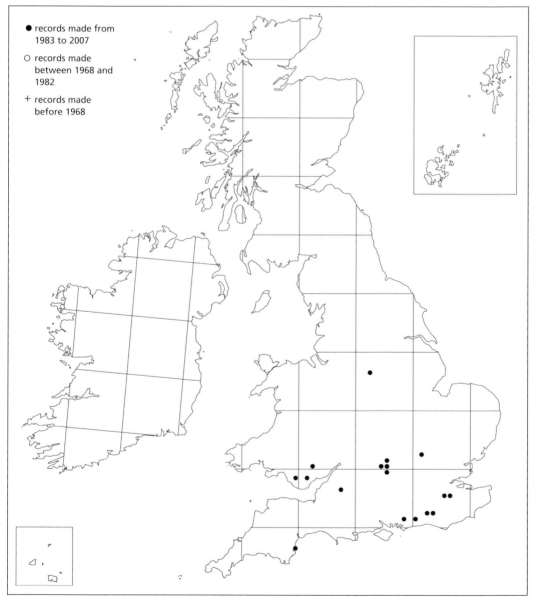

Map 18. *Haplophthalmus montivagus.*

Distribution: On current evidence *H. montivagus* is a rare woodlouse in Britain. The few records are mostly scattered across the chalk and limestone of south-eastern England. Recent records from Devon, south Wales and Derbyshire hint at a wider occurrence. The distribution reflects the activities of the few recorders that are able both to find and to reliably identify this elusive species. Although it undoubtedly remains under-recorded in many areas, it does seem to have a genuinely patchy distribution. For example, in Oxfordshire it occurs in good numbers at some sites (e.g. Wytham Woods, SP50), whilst at other apparently suitable woodlands (e.g. Wychwood Forest, SP31) *H. mengii* has been found instead (Gregory & Campbell, 1995).

Habitat: This species appears to be a strict calcicole; occurring in areas where friable soils have developed over limestone or chalk. In Oxfordshire *H. montivagus* also inhabits sites on heavy Gault clay (Gregory, 2001), which, although apparently unsuitable, is lime-rich and gives rise to friable soils during the dry summer months. Unlike *H. mengii*, *H. montivagus* has a quite narrow habitat preference. Over much of its range it typically inhabits deciduous, often ancient, woodland. It does occur in secondary woodland, for example in disused limestone quarries, but usually within well-wooded areas. It has been found in ornamental gardens in West Sussex (TQ23) and Derbyshire (SK26), but this is atypical. At the latter site, some 150km north of previously known localities, it is possible that the species was introduced through the importation of plants (Harper, 2004a).

Microsites: It is typically discovered clinging to the underside of deeply embedded stones and dead wood, but may be labouriously sorted from amongst friable humus-rich soil and limestone rubble. Although, primarily a soil-dwelling woodlouse it can be found within rotten dead wood. Hopkin and Roberts (1987) considered *H. montivagus* to favour wetter woodlands than *H. mengii*. Certainly, it is much easier to find in damp hollows or on stream edges, where it occurs closer to the surface.

Associated species: It is typically associated with other trichoniscid woodlice such as *H. danicus*, *Trichoniscus pygmaeus* and *Trichoniscoides albidus* and, like these species, it tends to be more easily found in the winter months.

Worldwide distribution: Occurring from Britain east to Poland and south to northern Italy, *H. montivagus* has a wide distribution across central European (Schmalfuss, 2004), but due to confusion with *H. mengii* it has been overlooked in many areas.

Metatrichoniscoides celticus Oliver & Trew, 1981

Distinctive features: This is a small slow-moving species, reaching 2.5mm. Live specimens of *M. celticus* are pure white, entirely lacking ocelli and the dorsal surface is covered in coarse tubercles. It is virtually identical in appearance to *M. leydigii* and to preserved specimens of *Trichoniscoides* (in which body and eye pigments are lost). Reliable identification can only be made from a male specimen.

Distribution: First collected from several sites along the coast of Glamorgan, south Wales, in 1979, *M. celticus* was described new to science by Oliver and Trew (1981). In March 1986 the species was also collected from a disused limestone quarry at Crwbin, Carmarthenshire, some 50km to the

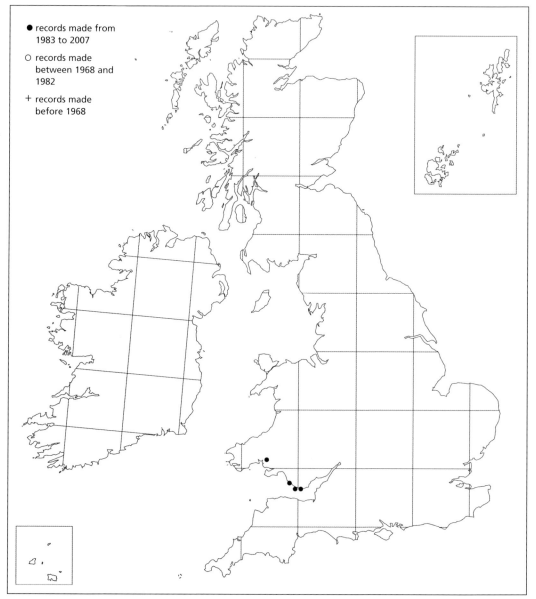

Map 19. *Metatrichoniscoides celticus.*

west (Chater, 1986b). Confirmed males are still only known from this region of south Wales. It is apparently endemic to the British Isles, being unknown elsewhere in Europe.

Habitat: This species is primarily associated with sparsely vegetated erosion banks occurring above rocky shores. However, the Crwbin quarry site is 7km inland and at an altitude of 170m above sea level (Chater, 1986b), clearly indicating that M. *celticus* is not confined to the supralittoral zone.

Female specimens, believed to be M. *celticus*, have been collected from St Bees Head, Cumberland (NX91) and from the Giant's Causeway in Co. Antrim, Ireland (C94). Since other species, such as M. *leydigii*, occur on the coasts of Europe these two records have not been mapped.

Microsites: Typically specimens are found in the supralittoral zone under large stones firmly embedded into damp humus-rich soil. Characteristically, small numbers of specimens are found with extreme difficulty. At Crwbin specimens were found under large blocks of limestone partially embedded into damp stony soil (Chater, 1986b).

The female specimen from St Bees Head was hand sorted from among accumulations of damp old red sandstone rubble at the base of the sea cliffs (Hopkin, 1987b). Subsequent visits to this site have failed to provide additional specimens. That at the Giant's Causeway was found under a stone embedded in grass turf growing on sand derived from basalt (Irwin, 1992).

Associated species: M. *celticus* is typically associated with other trichoniscids, usually *Haplophthalmus mengii*, *Trichoniscoides saeroeensis* and *Trichoniscus pygmaeus*.

Other notes: This is clearly an exceptionally elusive species and, even at known localities, its occurrence is unpredictable. Since the southern Welsh coast has been particularly well recorded it is likely that M. *celticus* may occur more widely around the less well-known coasts of western Britain and Ireland. For this reason, it is listed in the British Red Data Book (Bratton, 1991) as 'Insufficiently Known'. Considering its strong association with semi-natural coastal habitats, this species is potentially vulnerable to marine pollution, human disturbance and development of coastal sites (Harding & Sutton, 1985; Bratton, 1991). On the grounds of its habitat preference, apparent rarity, vulnerability and endemic status M. *celticus* was submitted for inclusion as a UK Priority BAP Species. However, this application was rejected because Continental Europe has been much less thoroughly surveyed than Britain and it is possible that M. *celticus* has been overlooked there (and, therefore, would not be endemic to Britain).

Worldwide distribution: The genus *Metatrichoniscoides* originated in the Atlantic region of Europe (Vandel, 1960). M. *celticus* may have evolved on the Atlantic coasts of Europe and spread north to Britain after the end of the last Ice Age (c. 10,000 years BP). In which case it may still be present on the Continent. Alternately, it may be autochthonous in Britain, having evolved following isolation from mainland Europe, in which case it is truly endemic.

Metatrichoniscoides leydigii (Weber, 1880)
Metatrichoniscoides leydigi (Weber, 1880)

Distinctive features: This species is virtually identical in appearance to M. *celticus* and to preserved specimens of *Trichoniscoides* (in which body and eye pigments are lost). M. *leydigii* can only be reliably identified from a male specimen.

Distribution: M. *leydigii* was first discovered in Britain in 1989 in the grounds of a garden centre in Oxford city (Gregory & Campbell, 1995). This remains the only known British locality for this species.

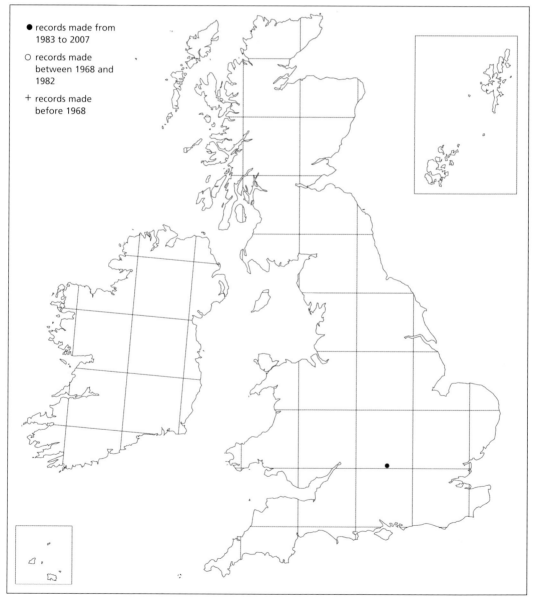

Map 20. *Metatrichoniscoides leydigii.*

Figure 11. *Metatrichoniscoides leydigii*. © Theodoor Heijerman.

Habitat: The garden centre lies in suburban Oxford on the site of a former plant nursery established over a century previously.

Microsites: Throughout the winter of 1989/1990 a thriving population was discovered outdoors within a discreet area of made-up ground composed of around ½ metre depth of ballast mixed with sand and topped with gravel. Peaty compost washed from plant pots had accumulated within the surface layers of gravel. Specimens of M. *leydigii*, associated with *Haplophthalmus mengii* and *Trichoniscus pygmaeus*, were seen beneath wooden pallets holding potted plants, beneath large plant pots (containing trees), within compost inside plant pots and amongst underlying peaty gravel. During the following summer M. *leydigii* became very elusive, but could be hand sorted from amongst sand and ballast at a depth of about 30cm.

Other notes: The Oxford population is undoubtedly an accidental introduction and may have been present, unnoticed, for many decades (Gregory, 2001). This site, which also supports *Buddelundiella cataractae*, has been redeveloped for housing and it will be interesting to see whether M. *leydigii* survives. Hopkin (1990a) suggests that M. *leydigii* may have been introduced from this locality, via potted trees, into gardens across Oxfordshire. Considering the unique nature of the microsites where it was found, this is unlikely. M. *leydigii* is widespread in the low-lying Netherlands (Berg, Soesbergen, Tempelman & Wijnhoven, 2008) and may prove to be more widespread in Britain. Although well-established garden centres, plant nurseries and botanic gardens are likely places to search, it is possible that M. *leydigii* has been overlooked in low-lying coastal areas of eastern England.

Worldwide distribution: M. *leydigii* occurs in coastal areas of north-western Europe from the Atlantic coast of France east to Belgium and western Germany. It is known as an introduction in glasshouses in Czechia, Sweden and Finland (Schmalfuss, 2004).

Miktoniscus patiencei Vandel, 1946

Distinctive features: M. *patiencei* is a small pure white woodlouse, reaching 4mm. The eyes consist of a single prominent black ocellus, its colour retained in alcohol, and the entire body is covered in coarse spiny tubercles. *Trichoniscoides saeroeensis* is similar, but has reddish/orange ocelli, which rapidly lose their colour in alcohol, and the dorsal surface is less coarsely tuberculate.

Distribution: First recorded in Britain as recently as 1976 (Oliver & Sutton, 1982), our understanding of this species' distribution has improved significantly since the publication of Harding and Sutton (1985). It occurs widely along the southern coasts of England from Lundy in the Bristol Channel to Aldeburgh in Suffolks. Outlying populations are known from Anglesey in

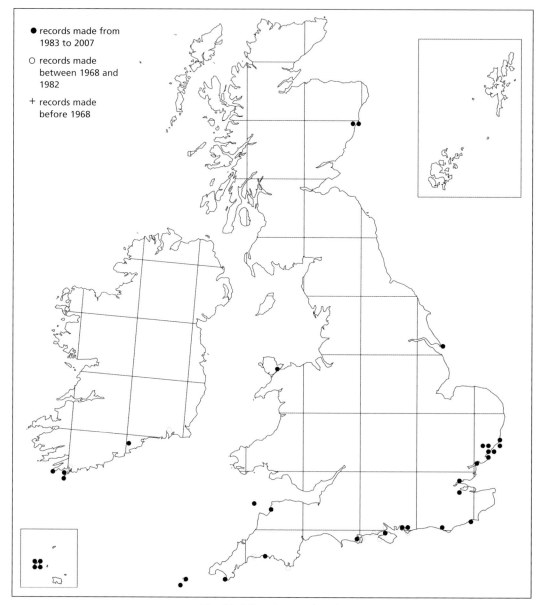

● records made from
    1983 to 2007

○ records made
    between 1968 and
    1982

+ records made
    before 1968

Map 21. *Miktoniscus patiencei.*

Figure 12. *Miktoniscus patiencei.* © Theodoor Heijerman.

North Wales, Spurn Head in Yorkshire and two localities on the Kincardineshire coast of eastern Scotland. In Ireland it appears to be restricted to the southern coast where it is distinctly local (M. Cawley, personal communication). Although it has not been found at many apparently suitable localities, the outlying localities hint at a much wider distribution around the British and Irish coastline.

Habitat: This is an exclusively coastal species primarily associated with supralittoral habitats. It should be sought in salt marsh, sparsely vegetated shingle banks, erosion banks and at the base of sea cliffs, often associated with pockets of damp friable peaty soil. On sea cliffs it has been found up to 40 metres above the level of extreme high tides.

Microsites: It can be found among grass roots, grass litter and strandline debris or beneath stones embedded into maritime turf. In sparsely vegetated shingle it typically occurs several centimetres down where damp peaty soil derived from rotted strandline material has accumulated. The species can be elusive at some sites, but it may be abundant in favoured microsites.

Associated species: It is frequently found associated with other soil-dwelling trichoniscids, such as *Trichoniscoides saeroeensis*, *Trichoniscus pygmaeus* and *T. pusillus* agg.

Worldwide distribution: M. *patiencei* has a restricted Atlantic distribution from Britain and Ireland, along the northern coast of France (Schmalfuss, 2004) and reaching at least as far south as north-western Spain (Gregory, 2004).

Oritoniscus flavus (Budde-Lund, 1906)

Distinctive features: This is a medium sized species, reaching 8mm and is capable of rapid movement. It is reminiscent of a large, dark *Trichoniscus pusillus*, but the eye is composed of a single prominent ocellus and body pigments fade in alcohol (*Trichoniscus* has three ocelli and pigments are retained).

Distribution: O. *flavus* is virtually confined to Ireland and is locally common across many counties in the south east, predominantly within the catchments of the rivers Barrow, Nore and Suir. The northernmost populations occur in Co. Galway on the west coast and Co. Meath on the east coast. For many years it was a mystery as to why this 'biogeographical curiosity' did not occur in Britain (Harding & Sutton, 1985). However, in March 1994 a population was discovered south east of Llanelli in Carmarthenshire, south Wales (Morgan, 1994). No additional British localities have been discovered.

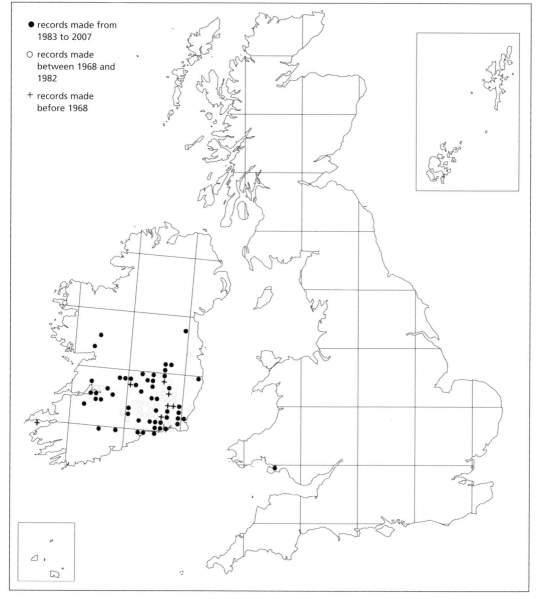

Map 22. *Oritoniscus flavus.*

Figure 13. *Oritoniscus flavus*. © Paul Richards.

Habitat: In Ireland it is characteristically associated with river and stream valleys, often occurring within rough vegetation beside watercourses. Occasionally, it has been found in coastal habitats, including salt marsh strandline (Doogue & Harding, 1982) and coastal shingle (Cawley, 2001). Although recorded from waste ground by Cawley (2001), it generally avoids synanthropic sites. The Carmarthenshire locality is low-lying coastal pasture (< 5m asl), reclaimed from saltmarsh early in the 19th Century. Further fieldwork has shown that permanently damp, rank grassland, scrub and marshland without any saline influence is favoured (J.F. Harper & I.K. Morgan, personal communication).

Microsites: In Ireland it can be found among leaf litter, under stones and dead wood, sometimes at the bases of walls and bridges. The initial Welsh specimens were collected from waterlogged red-rotted timber lying in a grassy drainage ditch (Morgan, 1994; Morgan & Pryce, 1995). Subsequently, specimens have been collected from the surrounding area from among rich dark humic soil or beneath stones.

Other notes: The core of the Irish populations is centred on an area known to support other thermophilous woodland/grassland species (P.T. Harding, personal communication). This distribution pattern and the species strong association with semi-natural habitats has been taken as evidence of its native status in Ireland (Doogue & Harding, 1982). In contrast, Cawley (2001) considers that it may be an ancient introduction that has become completely naturalised in damp habitats in Ireland and that it is possibly expanding north-westwards.

The Welsh population is likely to have been introduced. During the 19th Century coal from the Llanelli region was exported to many localities, including Ireland and the western seaboards of France and Spain (Morgan & Pryce, 1995). O. *flavus* might have arrived with ballast from returning ships, as have many plants introduced to the area. Specimens from Carmarthenshire have been confirmed as O. *flavus* by Prof. H. Daelens. These have retained much of their original body pigmentation in alcohol (P.T. Harding, personal communication), which is quite different to the rapid loss of pigment typically seen in Irish specimens (Gregory, 2003). One possible explanation of this different behaviour is that O. *flavus* might have been introduced to Wales from continental (rather than Irish) stock.

Worldwide distribution: O. *flavus* has a strict Atlantic distribution, and in addition to Ireland, occurs in north-eastern Spain and south-western France (Schmalfuss, 2004). Co. Meath, Ireland, represents the most northern locality in the world.

Trichoniscoides albidus (Budde-Lund, 1880)

Distinctive features: This small woodlouse of up to 4mm is similar in size, shape and colour to the ubiquitous *Trichoniscus pusillus* agg. Being slow moving it often remains after the *T. pusillus* agg. specimens have dispersed to seek shelter. The coarsely tuberculate body surface, the single ocelli and rapid loss of pigment in alcohol differentiate *T. albidus* from *T. pusillus* agg..

Distribution: Our knowledge of this species' distribution has improved significantly since the publication of Harding and Sutton (1985). In southern and eastern Britain it has proved to be widespread, but it is genuinely rare west of a line from Dorset to Lincolnshire. To illustrate this point, Jon Daws (1994a) found this species with difficulty at a single Leicestershire locality, but subsequently found it widely in Suffolk (Daws, 1995b). An outlying population is known from the

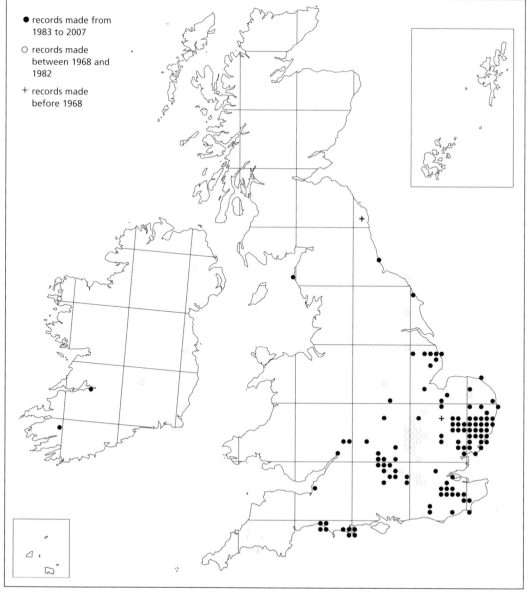

● records made from
  1983 to 2007

○ records made
  between 1968 and
  1982

+ records made
  before 1968

Map 23. *Trichoniscoides albidus*.

Figure 14. *Trichoniscoides albidus.* © Paul Richards.

north-eastern coast at St Bees Head, Cumberland (Hopkin, 1987b). Sporadic sites are known across southern Ireland where the species may be genuinely rare. The patchy distribution of *T. albidus* is primarily because, in the field, most recorders do not recognize this species amongst the background of *T. pusillus* agg.. This map, more than any other, illustrates areas where experienced recorders have been active. It has been severely under-recorded in many areas, including Ireland (Cawley, 2001).

Habitat: This is a soil-dwelling species with a strong preference for damp friable soils, but not usually waterlogged soils. An association with alluvial soils is noted in Harding and Sutton (1985) and many of the localities towards the north and west of its range are associated with river catchments. In the Thames Valley of Oxfordshire *T. albidus* is common (albeit elusive), but it is rarely encountered elsewhere in the county (Gregory & Campbell, 1995). Typical habitats include wet deciduous woodland and alluvial meadows, but it can occupy a broad range of habitats, often near watercourses, and on the coast, seepages on 'soft' slumping cliffs. It is often been found associated with 'dry' road-side ditches and occupies synanthropic sites such as churchyards in Suffolk (Daws, 1995b) and farmyards in Lincolnshire (Daws, 1993c). In Ireland it has mostly been recorded from synanthropic sites, such as gardens (Doogue & Harding, 1982) and churchyards (Gregory, 2002a), but it has also been collected from an estuarine erosion bank in Co. Limerick (Cawley, 2001).

Microsites: It is rarely found at the soil surface and specimens, usually singletons, are typically found with difficulty on the underside of embedded stones and dead wood or amongst rubble. However, *T. albidus* does seem to occur closer to the surface than other *Trichoniscoides* spp. and may be sorted from moist leaf-litter and superficial layers of damp friable soil. In Oxfordshire it has occasionally been taken in pitfall traps (Gregory & Campbell, 1995). In the dry summer months it can be very elusive, but like many trichoniscids it is much easier to find in winter. Wet or frosty conditions (Daws, 1995b) and sunny days following prolonged frosty spells can be most productive.

Associated species: It is typically found with *T. pusillus* agg., *T. pygmaeus* and *Haplophthalmus* spp.

Worldwide distribution: *T. albidus* has an Atlantic distribution, occurring across north-western Europe, from southern Sweden and Denmark through the Netherlands, Belgium, Britain and Ireland to western France (Schmalfuss, 2004).

Trichoniscoides saeroeensis Lohmander, 1923

Distinctive features: This is a small, to 4mm, creamy-white woodlouse flushed with varying amounts of pinkish-orange. The dorsal surface is covered with small tubercles and the eye is composed of a single pinkish-red ocellus. In life, and when preserved in alcohol, it is very similar in appearance to *T. helveticus* and *T. sarsi* and, ideally, identification should be based on a male specimen.

Distribution: As predicted by Harding and Sutton (1985), the map indicates a widespread distribution around the entire coast of Britain and Ireland from the Isles of Scilly in the south to the Shetland Islands in the north.

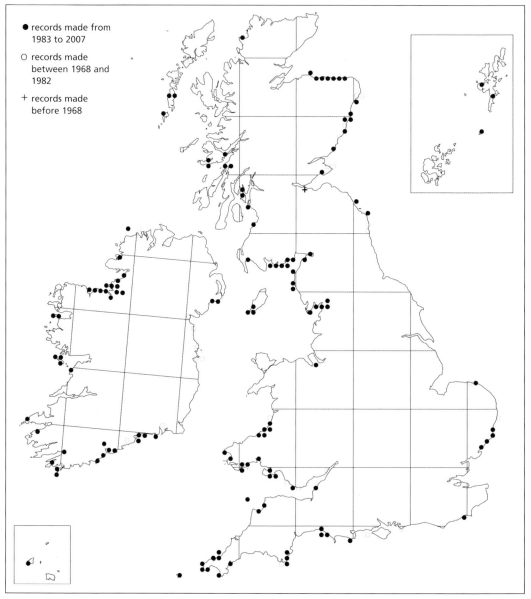

● records made from
   1983 to 2007

○ records made
   between 1968 and
   1982

+ records made
   before 1968

Map 24. *Trichoniscoides saeroeensis*.

Habitat: Although most characteristic of coastal erosion banks, this species is found in a variety of sheltered coastal habitats, including sparsely vegetated shingle and salt marsh strandline. It is usually found in the supralittoral zone, but may penetrate up to 100 metres inland where sparse maritime grassland occurs on unstable cliffs. Although predominately coastal, this species is clearly capable of surviving away from maritime influences. The earliest British and Irish records for *T. saeroeensis* were subterranean (Sheppard, 1968) and it has proved to be a frequent troglophile and can occur deep within caves and mines.

More recently *T. saeroeensis* has been found inhabiting the summits of limestone mountains. Daws (1993b) records it from limestone pavement at Hutton Roof Crags and Newbiggin Crags, Cumbria (at 200m asl and 10km inland). In Ireland, Cawley (1996) found it to be common on summits of the Dartry range in Counties Sligo and Leitrim (350-450 m asl and 12km inland). The presence of *T. saeroeensis* in the uncultivated limestone uplands of England and Ireland has all the indications of a natural occurrence. Moseley (1995) suggests that in limestone areas it may inhabit subsurface mesocavernous voids and that the flow of interstitial waters carries it into underlying cave and mine systems. Competition from other woodlice, such as *Trichoniscus pygmaeus*, may restrict its distribution to either exposed coastal sites or equally exposed hill-top locations (Cawley, 1993).

Microsites: On the coast, *T. saeroeensis* is usually found several centimetres below the surface layers where damp peaty soil has accumulated amongst shingle or on the underside of large stones partly embedded into friable soil. Often it can be found on the taproots of plants, such as sea-kale, *Crambe maritima*. At its upland Cumbrian localities it was found under stones and rocks partially embedded into soil (Daws, 1993b). Specimens from the Dartry range were found within damp peaty soil where rocky ground, with sparse relict Alpine flora, broke through blanket bog (Cawley, 1996). In caves and mines it is often found within waterlogged timbers (Moseley, 1995). It is typically found in small numbers and, in common with many trichoniscids, the ease of locating this species depends upon current soil and weather conditions.

Worldwide distribution: *T. saeroeensis* is restricted to the coastline of north-western Europe, from northern France, around Britain and Ireland north to Denmark and southern Sweden (Schmalfuss, 2004). Interestingly, it is absent from the coastline of the Netherlands, where *T. sarsi* occurs instead (Berg, Soesbergen, Tempelman & Wijnhoven, 2008).

Trichoniscoides sarsi aggregate
Trichoniscoides sarsi sensu lato

Taxonomic note: The *T. sarsi* of Harding and Sutton (1985) is now known to consist of two closely allied species *T. sarsi* Patience and *T. helveticus* (Carl) (Hopkin, 1990b). Both species, particularly when preserved in alcohol, are identical in appearance and the two can only be separated by microscopic examination of a male specimen. In the absence of a male specimen, records for the two species are conveniently lumped together as the *T. sarsi* aggregate. The map shows records where female specimens have been collected, or where recorders have not dissected a male specimen.

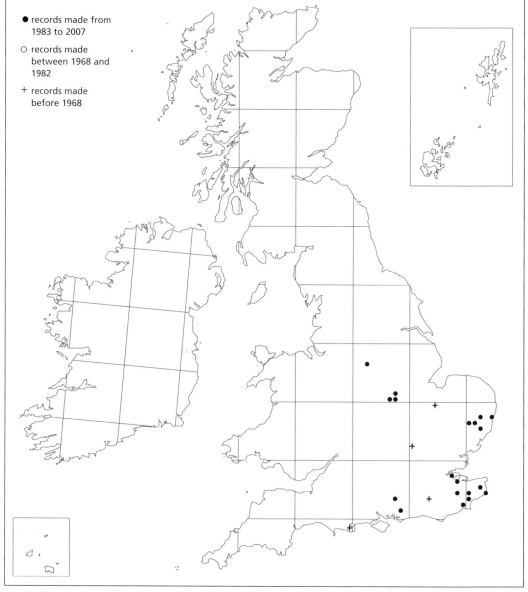

Map 25. *Trichoniscoides sarsi aggregate.*

Distribution: The records are in keeping with the known ranges of the two species; from Dorset north to Derbyshire and east to Suffolk in Britain and clustered around Dublin in Ireland. Although on current evidence it appears that *T. sarsi* and *T. helveticus* have mutually exclusive distributions in Britain and Ireland, the records plotted here, and those in Harding and Sutton (1985), could refer to either species.

Trichoniscoides helveticus (Carl, 1908)
Trichoniscoides helveticus Carl, 1908

Taxonomic note: *T. helveticus* was first discovered in Britain in 1990 when Steve Hopkin (1990b) demonstrated that most English specimens, and many published records, of *T. sarsi* Patience were actually referable to *T. helveticus* (Carl). Thus, the species was unknown to Harding and Sutton (1985) and the few records were inadvertently included within the map of *T. sarsi*. The two species can only be separated by examination of a male specimen (hence the past confusion) and undifferentiated records are lumped together as the *T. sarsi* aggregate.

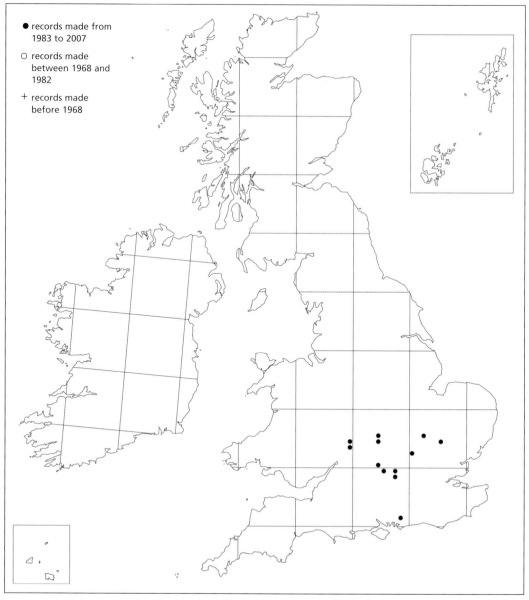

● records made from
   1983 to 2007

○ records made
   between 1968 and
   1982

+ records made
   before 1968

Map 26. *Trichoniscoides helveticus*.

Figure 15. *Trichoniscoides helveticus.* © Theodoor Heijerman.

Distinctive features: Another small creamy-white woodlouse flushed with varying amounts of pinkish-orange. The dorsal surface is covered with distinct tubercles and the eye is composed of a single dark reddish ocellus. All pigmentation is lost in alcohol. *T. helveticus* is identical in appearance to *T. sarsi* and only males can be identified.

Distribution: Scattered localities are known from across central southern England. It has proved to be widespread in the well-surveyed county of Oxfordshire (Gregory & Campbell, 1995). Other records are from Worcestershire, Buckinghamshire, Bedfordshire, Huntingdonshire, North Essex and West Sussex. There are no confirmed records from Ireland (where *T. sarsi* has been recorded). It is likely to be considerably under-recorded and could be expected to turn up in appropriate habitats anywhere in southern England.

On current evidence the British ranges of *T. helveticus* and *T. sarsi* do not overlap. This cannot be explained by differences in habitat preferences or recorder bias (Gregory, 2002b). The two species are also mutually exclusive in the Netherlands where their distributions are strongly tied to the underlying geology. *T. helveticus* occupies eastern areas where Pleistocene 'river' clays occur above sea level, but Berg (2008) suggests that other factors, such as climate, are also important in explaining the observed distribution. There is likely to be a similar geological and climatic explanation for the pattern seen in Britain and further records, with associated habitat data, may help elucidate this puzzle.

Habitat: Unlike *T. sarsi*, this species shows a strong preference for semi-natural habitats, suggesting that it may be a native species. However, it is not dependent on any particular vegetation type and has been collected from species-rich short-turf grassland, rank *Arrhenatherum* grassland, mixed scrub and deciduous woodland. In Oxfordshire it has also been taken from amongst limestone rubble beside a man-made reservoir. It would appear to be restricted to sites with undisturbed friable chalk or limestone soils and these tend to be associated with semi-natural grasslands and woodlands.

Microsites: Specimens are typically found beneath large stones partially embedded into soil, but it can be sorted from amongst chalky soil or limestone rubble. It is very elusive and often a single specimen is found with difficulty. A recurrent theme is that *T. helveticus* is much easier to find during inclement weather conditions and remains active even in heavy frosts. In summer it moves deeper into the soil and becomes much more elusive.

Associated species: It is often associated with other soil-dwelling trichoniscids, primarily *Trichoniscus pygmaeus*.

Worldwide distribution: This species, believed to have originated in the western Alps (Vandel, 1960), is known from Switzerland, Germany, Czechia, western France, Belgium, the Netherlands and England (Schmalfuss, 2004).

Trichoniscoides sarsi Patience, 1908
Trichoniscoides sarsi seg.
Trichoniscoides sarsi sensu stricto

Taxonomic note: Hopkin (1990b) has shown that the *T. sarsi* mapped in Harding and Sutton (1985) refers to two different species: the true *T. sarsi* Patience detailed here and *T. helveticus* (Carl). British and Irish records for *T. sarsi* published before 1990 could refer to either species.

Distinctive features: *T. sarsi* is identical in size and appearance to *T. helveticus* and can only be identified by microscopic examination of a male specimen, hence the past confusion.

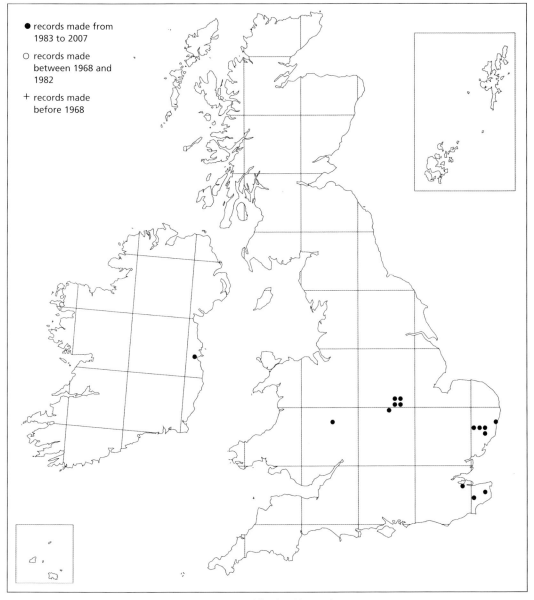

● records made from
  1983 to 2007

○ records made
  between 1968 and
  1982

+ records made
  before 1968

Map 27. *Trichoniscoides sarsi*.

Figure 16. *Trichoniscoides sarsi.* © Paul Richards.

Distribution: The distribution of *T. sarsi* occupies a distinct band across eastern and central England and into eastern Ireland. Hopkin (1990c) reports male specimens from Co. Dublin, Leicestershire and Kent. Subsequently, it has proved to be widespread in Leicestershire and Suffolk (Daws, 1994a, 1995b) and it has recently been recorded in Shropshire (J.P. Richards, personal communication). Cawley (2001) re-found the species in Dublin.

It is apparent that the majority of the records are attributable to the few experienced recorders, notably Jon Daws, who are capable of finding this species. The conclusion is that *T. sarsi* is very elusive and extremely under-recorded. However, the apparent mutually exclusive distribution with *T. helveticus* does seem to be genuine. For example, thorough surveys of churchyards in Oxfordshire, an area where *T. helveticus* is well known, have failed to discover *T. sarsi* (Gregory & Campbell, 1995). In the Netherlands, where the 'coastal' *T. saeroeensis* has not been recorded, *T. sarsi* occurs widely on the coast (Berg, Soesbergen, Tempelman & Wijnhoven, 2008). This raises the possibility that *T. sarsi* has been overlooked on the eastern coasts of Britain. It almost certainly occurs elsewhere and should be sought in synanthropic situations, at least within its known range across England and Ireland.

Habitat: Characteristically, this is an urban species associated with disturbed sites. Synanthropic sites, such as old gardens or churchyards in the environs of towns and villages, are quite typical. Its preference for disturbed sites suggests that *T. sarsi* is a well-established introduction.

Microsites: Specimens are usually found beneath 'large stones', such as paving slabs, fallen gravestones and chunks of concrete, partially embedded into the soil. Additionally, *T. sarsi* can be sorted from soil or amongst rubble, often at the base of walls. It is elusive and rarely found in large numbers, typically just one or two specimens. Like many soil-dwelling trichoniscids, it is much easier to find in winter, particularly during heavy frost (Doogue & Harding, 1982; Daws, 1995b).

Associated species: It often occurs with other trichoniscids, such as *Trichoniscus pygmaeus*, *Haplophthalmus mengii* and, in Suffolk, *Trichoniscoides albidus*.

Worldwide distribution: *T. sarsi* has an Atlantic distribution and is also known from northern France, western Germany, Netherlands, Denmark, southern Norway and southern Sweden (Schmalfuss, 2004). Vandel (1960) believes it originated in western France and has dispersed northwards along the coastline to other western European countries.

Trichoniscus pusillus aggregate

Trichoniscus pusillus sensu lato

Trichoniscus pusillus Brandt (of many earlier works)

Trichoniscus pusillus ff (of BRC card RA 51)

Taxonomic note: This is the *T. pusillus* Brandt mapped and discussed in Harding and Sutton (1985). In Britain and Ireland *T. pusillus* is known to occur as two distinct races, the sexual *provisorius* (male/female ratio is 1:1) and the parthenogenetic *pusillus* (males virtually absent). Although, traditionally treated as sub-species (Vandel, 1960; Gruner, 1966) they cannot interbreed. In keeping with other species with parthenogenetic races, such as the bristly millipede *Polyxenus lagurus* (Enghoff, 1976), Harding and Sutton (1985) treated the two races as forms. Schmalfuss (2004) elevates

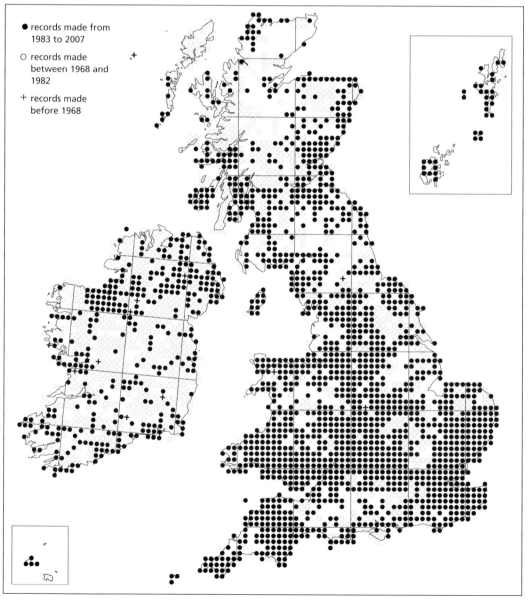

● records made from
1983 to 2007

○ records made
between 1968 and
1982

+ records made
before 1968

Map 28. *Trichoniscus pusillus* aggregate.

Figure 17. *Trichoniscus pusillus* aggregate. © Theodoor Heijerman.

them to the status of full species. The great majority of recorders have not differentiated between these two species and they are conveniently lumped together as *T. pusillus* agg. and plotted here.

Distinctive features: *T. pusillus* agg. reaches 5mm in length and is the only common small woodlouse that is mottled reddish-brown. The eyes are composed of three fused ocelli and body surface is smooth. Being morphologically identical, the two species can only be reliably separated by microscopic examination of male genitalia or by detailed examination of the ratio of males to females within discreet populations (Fussey, 1984). Unfortunately, both species frequently occur in close proximity. Frankel, Sutton and Fussey (1981) indicate that the proportion of *T. pusillus* to *T. provisorius* at any given site can be estimated by measuring the ratio of males to females. However, the identification of female specimens is seldom possible.

Distribution: Recorded from all vice-counties in England, Wales, Scotland and Ireland, *T. pusillus* aggregate is probably our most numerous woodlouse. Densities of several 1000 per square metre have been recorded in deep grass litter. During casual surveys it is easily overlooked because of its small size and is only our third most frequently recorded woodlouse (after *Oniscus asellus* and *Porcellio scaber*). Detailed surveys undertaken in the 1970s indicate that both species within the aggregate are geographically equally widespread throughout Britain and Ireland. They are able to co-exist throughout this range through their ability to exploit subtly different niches within any given locality (Fussey, 1984).

Habitat: Being tolerant of high levels of natural and human disturbance, acidic substrates and high altitude, this species aggregate is ubiquitous in a great variety of habitats provided that sufficient moisture is present. Favoured sites include supralittoral coastal sites, marshes, grasslands, woodlands, gardens and a great variety of synanthropic sites. In upland areas it is frequent along access roads, beside watercourses and on damp ledges. It may be the most frequent woodlouse in Irish upland oak woodland (Doogue & Harding, 1982) and even penetrates blanket bog in the west (Cawley, 1996). Above the tree-line of the Scottish highlands, where it is associated with herb-rich ledges, it occurs to 800m asl and may be the only woodlouse present (G.B. Corbet, personal communication).

Microsites: It is usually found close to the soil surface, typically on the underside of stones and dead wood or sorted from amongst leaf-litter and moss. In dry weather it moves deeper into soil and becomes more elusive.

Worldwide distribution: This species aggregate is very widespread throughout Europe, from the Iberian peninsula in the west, north to Iceland and Scandinavia, east to Turkey and south to Algeria and widely introduced elsewhere.

71

Trichoniscus provisorius Racovitza, 1908
Trichoniscus pusillus form *provisorius* Racovitza, 1908
Trichoniscus pusillus ssp. *provisorius* Racovitza, 1908

Taxonomic note: This is the *T. pusillus* form *provisorius* Racovitza mapped in Harding and Sutton (1985). Schmalfuss (2004) elevates this to full species status, which essentially adds an additional species, *T. provisorius* Racovitza, 1908, to the British and Irish check-list (Gregory, 2006).

Distinctive features: *T. provisorius* is a normal diploid, sexually reproducing species with populations composed of approximately equal numbers of males and females (Vandel, 1960; Gruner, 1966; Frankel, Sutton & Fussey, 1981). Morphologically it is identical to *T. pusillus sensu stricto*, but is

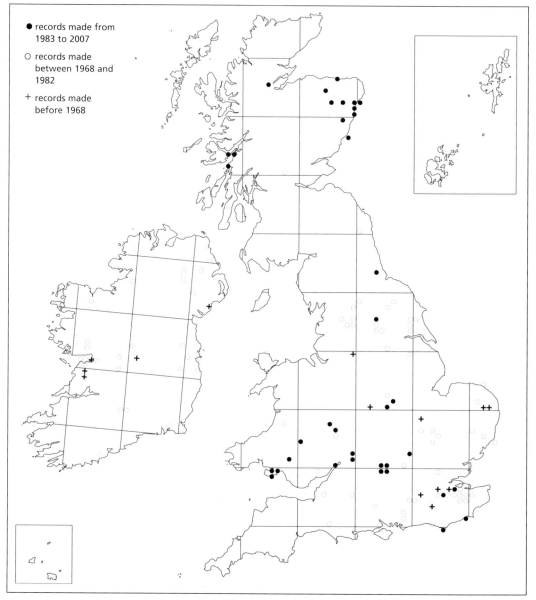

● records made from
   1983 to 2007

○ records made
   between 1968 and
   1982

+ records made
   before 1968

Map 29. *Trichoniscus provisorius*.

Figure 18. *Trichoniscus provisorius.* © Paul Richards.

comparatively smaller at each stadium and specimens rarely exceed 3.5mm in length when fully grown. Males can be reliably separated from those of *T. pusillus sensu stricto* by microscopic examination of the first exopod.

Distribution: As with *T. pusillus sensu stricto*, this species shows a wide distribution across Britain and Ireland. Although confirmed records occur as far north as Banff and East Ross, northern Scotland, there is a noticeable bias to south-eastern England. Fussey (1984) showed this to be a result of the predominance of calcareous strata in the south-east and that *T. provisorius* was also common in other major limestone areas, such as the Carboniferous strata of northern England and the Burren in Ireland.

Habitat: Although the two species frequently coexist at any given locality it is apparent that *T. provisorius* favours more freely draining and sparsely vegetated calcareous localities, such as south-facing rocky slopes or limestone pavement (Fussey, 1984). However, *T. provisorius* is not a strict calciphile. It favours high levels of insolation and benefits from the ability of limestone tracts to gain heat rapidly and from their tendency to produce free draining soils. In Oxfordshire, *T. provisorius* has been collected in sunny situations, such as semi-natural limestone grassland and a herb garden, but also from free-draining woodland.

Microsites: It is generally found in similar places to *T. pusillus*, such as beneath dead wood, under stones, including paving slabs, and at the base of walls, but possibly in sunnier and/or drier situations.

Worldwide distribution: *T. provisorius* occurs widely across Europe, including the Mediterranean basin, as far east as Turkey and the Lebanon and as far south as Algeria (Schmalfuss, 2004). It has been introduced to the Azores, Hawaii and North America.

Trichoniscus pusillus Brandt, 1833
Trichoniscus pusillus sensu stricto Brandt, 1833
Trichoniscus pusillus ssp. *pusillus* Brandt, 1833
Trichoniscus pusillus form *pusillus* Brandt, 1833

Taxonomic note: This is the *T. pusillus* form *pusillus* Brandt mapped in Harding and Sutton (1985). Schmalfuss (2004) elevates it to full species status, *T. pusillus* Brandt, 1833. This is a triploid, parthenogenetic species with populations almost entirely composed of females; males comprise about 1% of the population.

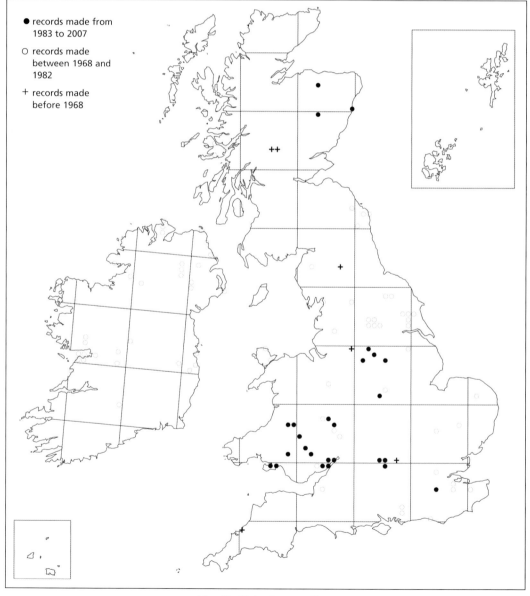

Map 30. *Trichoniscus pusillus.*

Distinctive features: Although identical in appearance to *T. provisorius*, it is consistently larger with specimens reaching 5mm. The two species can be reliably separated by microscopic examination of mature males. As males are rarely encountered, *T. pusillus sensu stricto* is relatively under-recorded. However, the virtual absence of males can be a convenient means of identification, but only if care is taken to sample discrete populations (Frankel, Sutton & Fussey, 1981; Fussey, 1984).

Distribution: *T. pusillus sensu stricto* is geographically widely distributed throughout Britain and Ireland. It becomes relatively scarce in south-eastern counties of England and Ireland where it seems to be partially replaced by *T. provisorius*. Confirmed records extend north to the Isle of Mull in the west and Banff in the east.

Habitat: This species has a very broad habitat niche. There is considerable overlap with *T. provisorius* and the two species frequently coexist at any given locality. However, *T. pusillus* appears to be better suited to cooler, damper habitats. For example, Fussey (1984) reports very low populations on limestone pavement, a habitat favoured by *T. provisorius*.

Microsites: It is usually found close to the soil surface, typically on the underside of stones and dead wood or amongst leaf-litter and moss. *T. pusillus* seems to occupy subtly different niches from *T. provisorius* (Fussey, 1984), generally favouring damper and/or shadier microsites.

Other notes: Being parthenogenetic and triploid typically makes a species more reproductively competitive, generally more vigorous and better adapted to withstand extreme conditions. This may explain why *T. pusillus* is typically the more numerous of the two species in most areas where discreet populations have been analysed (Fussey, 1984). It is likely that the majority of records for the species aggregate in northern and western areas refer to this species.

Worldwide distribution: Although widely dispersed across Europe, it mainly occurs west and north of the Alps and has been introduced into Madeira, the Azores and North America (Schmalfuss, 2004).

Trichoniscus pygmaeus Sars, 1898

Distinctive features: At 3mm this is one of our smallest woodlice. Colour is variable from off white, to pinkish or yellowish-apricot, but characteristically the head lacks pigmentation and appears relatively pale. Confusion is most likely with juvenile *T. pusillus* agg. in which the head is speckled with pigment to match the body and the three black ocelli, that comprise each eye, are not fused. *Trichoniscoides* spp. and juvenile *Androniscus dentiger* have the body covered in tubercles and have a single ocellus.

Distribution: The distribution map has considerably filled in since the publication of Harding and Sutton (1985). *T. pygmaeus* has proved to be a common woodlouse throughout much of Britain

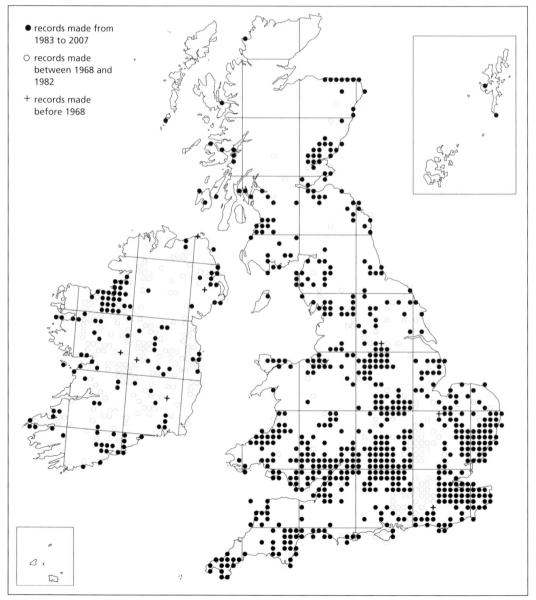

Map 31. *Trichoniscus pygmaeus.*

Figure 19. *Trichoniscus pygmaeus.* © Paul Richards.

and Ireland and should occur in most 10km grid squares both inland and on the coast. Although recorded from the Outer Hebrides and the Shetland Islands it does seem to become scarce in northern Scotland, but this may be a result of under-recording. The map shows much recorder bias and clearly indicates areas where key recorders have been particularly active.

Habitat: This soil-dwelling species can be found in most habitats, wherever suitable friable soil conditions occur. Although it can be found in semi-natural grasslands and woodlands, it is often much easier to find in synanthropic sites such as churchyards, disused quarries and railway cuttings. In coastal habitats this can be a common species in the supralittoral zone and above. Its habitats include sparsely vegetated shingle, erosion banks, boulder beaches, the base of sea cliffs and sparse maritime grassland on unstable cliffs or cliff tops.

Microsites: Searching the underside of large stones or dead wood partially embedded into soil, particularly in damp spots, is the simplest way to find this species. In the absence of these features, specimens can be hand sorted or sieved from soil. Within coastal shingle it is usually found several centimetres below ground where damp peaty soil has accumulated. It can also be found in caves, both inland and on the coast, but tends to occupy the cave threshold rather than areas deep inside cave systems (Moseley, 1995). During cold or wet weather it is much easier to locate, but is rarely found in large numbers, typically one or two specimens, a reflection of its elusive behaviour. With appropriate collecting methods up to 600 individuals per square metre have been recorded (Harding & Sutton, 1985).

Associated species: It is typically associated with other soil-dwelling trichoniscids, including *T. pusillus* agg., *Androniscus dentiger*, *Haplophthalmus* spp. and *Trichoniscoides* spp.

Worldwide distribution: This species has a wide distribution across Europe and extends into south-western Russia, the Azores and Morocco. It has been introduced into North America.

Halophiloscia couchii (Kinahan, 1858)
Halophiloscia couchi (Kinahan, 1858)

Distinctive features: This fast and agile species is rather reminiscent of an immature *Ligia oceanica*. The very long antennae with three elongated flagella segments are quite distinct from those of *L. oceanica* in which the antennal flagella are composed of numerous bead-like segments.

Distribution: Our understanding of this species' distribution has improved significantly since the publication of Harding and Sutton (1985). In the west it is known to occur around the entire coast of Wales with an outlying population at St Bees Head, Cumberland. It has proved to be widespread along the southern coast of England and extends eastwards as far as Ramsgate, Kent.

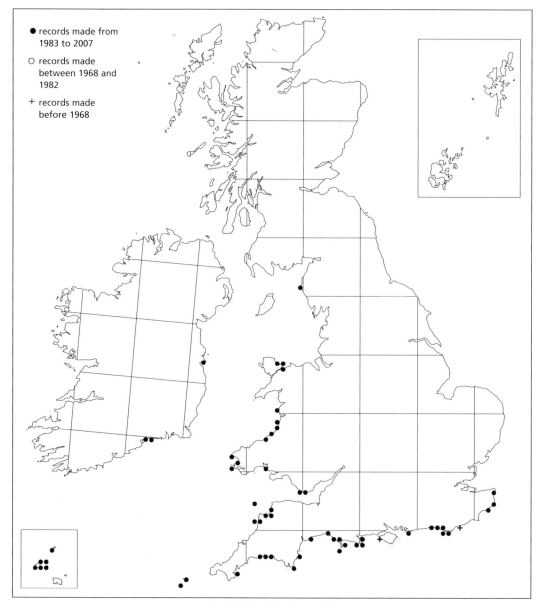

Map 32. *Halophiloscia couchii*.

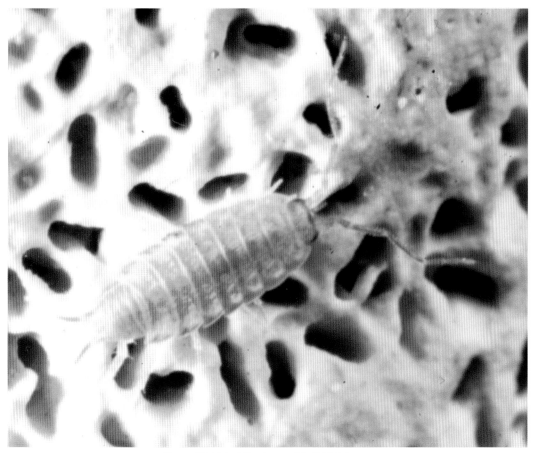

Figure 20. *Halophiloscia couchii*. © Dick Jones.

It has been recorded from many offshore islands including the Isles of Scilly (Jones & Pratley, 1987) and Lundy (Daws, 1991). In Ireland, *H. couchii* has been known from Howth Head, Co. Dublin since the early 20th Century, but it was not until 1998 that additional sites were discovered in Co. Waterford (Cawley, 2001).

H. couchii is probably under-recorded in many areas, particularly Ireland (Cawley, 2001). Given its south-western distribution in Britain, and the implication that it favours mild winters (Hopkin, 1987a), then its apparent scarcity in Ireland is surprising. It is worth searching for this species at the base of cliffs and on shingle beaches on the Isle of Man and the south-western coasts of Scotland.

Habitat: This is an exclusively coastal species rarely found far above the supralittoral zone. It is typically associated with unvegetated rocky cliffs, boulder beaches or shingle beaches. In Sussex it was found 100m inland, on an exposed low-lying vegetated boulder beach exposed to sea spray (Daws, 1993a), but this is unusual. By day it occupies dark, humid and inaccessible crevices and can be extremely difficult to locate. However, there is evidence that it can be easier to find at high tide, during wet weather or at night (e.g. Daws, 1992b; Hopkin, 1991b). Although Harding and Sutton (1985) considered it to be more plentiful at calcareous sites, many localities, such as those in west Wales (Chater, 1986b), are on acidic strata.

Microsites: An extremely active species, *H. couchii* rapidly seeks shelter when disturbed and it requires great agility to capture a specimen. It can be found by turning boulders or rocks on boulder storm terraces or at the base of sea cliffs. On rocky cliff faces it can often be found in holes and fissures packed with damp rotting seaweed (Daws, 1992b). On unvegetated shingle beaches this is a particularly difficult species to capture. According to Hopkin (1991b), the 'demented gerbil' method of squatting down whilst pushing large quantities of shingle between ones legs is a useful technique to pursue a specimen. It can be more easily sieved from shingle and Fowles (1989) records a case of it being sieved from rotting strandline material.

Associated species: In unvegetated shingle and boulder beaches it is often the only woodlouse to be found, but on rocky cliffs it typically occurs with *Ligia oceanica*.

Worldwide distribution: *H. couchii* occurs widely along the coasts of the Black Sea, the Mediterranean and the Atlantic coastline from Senegal, Africa to the British Isles (Schmalfuss, 2004). St Bees Head is the most northerly known locality in the world.

Stenophiloscia glarearum Verhoeff, 1908
Stenophiloscia zosterae Verhoeff, 1928

Distinctive features: This is a small species, reaching 6mm. Live specimens appear pinkish, somewhat reminiscent of a large pale *Trichoniscus pusillus* (Gregory, Whiteley & Wilde, 2001) or a pale immature *Porcellio scaber* (Daws, 1995a). The few British specimens observed in the field have tended to remain stationary or have moved slowly into a nearby crevice upon disturbance. This contradicts the opinion of Hopkin (1991a) and Oliver and Meechan (1993), based on the appearance of preserved specimens (i.e. slender with long legs), that S. *glarearum* may be capable of rapid movement when disturbed. This behaviour is very different to the superficially similar *Halophiloscia couchii*, which very rapidly seeks shelter when disturbed.

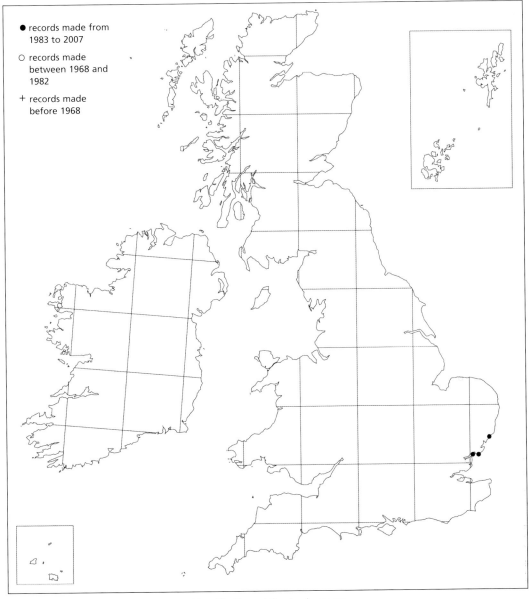

● records made from
  1983 to 2007

○ records made
  between 1968 and
  1982

+ records made
  before 1968

Map 33. *Stenophiloscia glarearum.*

Distribution: Apparently very rare in Britain, *S. glarearum* is known from a few sites along the Essex and Suffolk coastline and isolated records from Norfolk and Devon. The first British specimens were collected in pitfall-traps at Slapton Ley, Devon in 1974 and Scolt Head Island NNR, Norfolk in 1977. A live specimen was also collected by hand at Goldhanger, Essex in 1976 (Harding, Cotton & Rundle, 1980). Further pitfall trapping and hand searching had failed to produce additional material (Harding & Sutton, 1985). After an absence of nearly two decades a second live specimen was found at Shingle Street, Suffolk (Daws, 1995a) and subsequently in good numbers at Colne Point NNR in north Essex (Gregory *et al*, 2001).

Habitat: All specimens have been collected from between the high water mark and the storm drift line on unvegetated or sparsely vegetated shores composed of shingle or sand (Harding, Cotton & Rundle, 1980; Daws, 1995a; Gregory *et al*, 2001).

Microsites: The Shingle Street specimen was found under strandline driftwood on a sparsely vegetated shingle spit (Daws, 1995a). At Colne Point NNR specimens were found clinging beneath pieces of driftwood a few metres either side of the strandline along a several hundred metre stretch of sparsely vegetated sandy shingle beach (Gregory *et al*, 2001).

Associated species: Small numbers of *T. pusillus*, *Philoscia muscorum*, *P. scaber* and *Armadillidium album* were collected with *S. glarearum* at Colne Point, while the Shingle Street specimen was associated with abundant sand-hoppers (Decapoda).

Other notes: Although clearly not common, Harding and Sutton (1985) suggest that *S. glarearum* may be elusive rather than rare. The recent discovery at Colne Point, a well-surveyed site, supports this view. It is probable that the appearance of *S. glarearum* is triggered by a combination of factors, including weather and tides. On current evidence the species is widespread along the coast of East Anglia and could be expected to turn up on other undisturbed sandy shingle shorelines along the east and south coasts of England. Considering its specialist habitat, this species is potentially vulnerable to marine pollution and human disturbance.

Worldwide distribution: *S. glarearum* is widely dispersed along the Atlantic coastline of Europe and along the western Mediterranean coastline as far as Greece (Schmalfuss, 2004).

Philoscia muscorum (Scopoli, 1763)

Distinctive features: This medium sized species, reaching 11mm in length, runs rapidly when disturbed. Confusion is possible with *Ligidium hypnorum* and *Porcellionides cingendus*, which also have a stepped body outline and with which it may occur. In *P. muscorum* the antennal flagella are composed of three segments and there tends to be a distinctive dark dorsal stripe.

Distribution: This is predominantly a lowland species and over much of England and Wales it is best described as abundant and ubiquitous. In south-eastern counties, such as Oxfordshire (Gregory, 2001), *P. muscorum* is the most commonly encountered woodlouse. It becomes increasingly sparse in northern England and Scotland and becomes progressively more restricted to coastal sites and

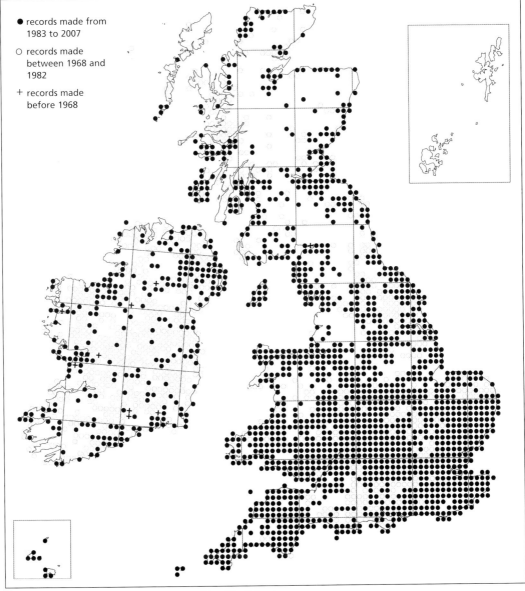

records made from 1983 to 2007

○ records made between 1968 and 1982

+ records made before 1968

Map 34. *Philoscia muscorum.*

Figure 21. *Philoscia muscorum*. © Theodoor Heijerman.

low-lying river valleys. It has not been recorded from the northern islands of Orkney and Shetland. In Ireland it is known from all vice-counties where it occurs widely and frequently.

Habitat: It has a strong preference for grassy sites, including rank grassland, species-rich short-turf, sand dunes, open scrub, hedgerows and gardens, but rarely penetrates far into shady woodland. In southern areas *P. muscorum* may also be found in fens and wet meadows and may coexist with the hygrophilous *L. hypnorum*. In gardens (and similar synanthropic localities) it is never as abundant as *Oniscus asellus* or *Porcellio scaber* and consequently is less commonly encountered by the general public. It favours calcareous conditions and tends to be much less frequent where acidic substrates occur. It is widespread in coastal habitats and in some coastal dune systems it may be the dominant woodlouse. However, in the south-western extremes of Ireland, England and Wales it seems to be partially replaced by *P. cingendus*. This is possibly due to unsuitable climatic conditions, such as relatively high humidity, rather than direct competition (Harding & Sutton, 1985).

In the more northern parts of its range *P. muscorum* favours dryish places, often on well-drained soils or where there is some shelter from inclement weather conditions. This includes swards, such as those composed of red fescue *Festuca rubra*, that dry out relatively quickly following rain or swards sheltered by sparse tree cover or on hedge banks. On higher ground it becomes restricted to sheltered ravines and grassy ledges. In western Scotland, where it is confined to the coast, *P. muscorum* is typically the last of the 'common' species to be found (after *P. scaber*, *O. asellus*, *Ligia oceanica* and *Trichoniscus pusillus* agg.) (G.M. Collis, personal communication).

Microsites: Although usually found under dead wood, stones or even dried cowpats, this undoubtedly reflects the relative difficulty of searching grassland swards and tussocks, where the species may also be found. In limestone areas it may be found amongst scree. In gardens and churchyards it is not uncommon in the compost heaps. It may be found under supralittoral vegetation and strandline debris.

Worldwide distribution: *P. muscorum* occurs widely in Europe as far north as southern Norway, east to Poland and south to central Greece (Schmalfuss, 2004).

Platyarthrus hoffmannseggii Brandt, 1833
Platyarthrus hoffmannseggi Brandt, 1833

Distinctive features: This is a small, blind, white woodlouse reaching 5mm in length. Its broad oval body, stout antennae and its association with ants are quite distinctive.

Distribution: Across much of southern Britain and southern Ireland *P. hoffmannseggii* occurs widely. South of a line from the Humber Estuary to the Severn Estuary, and in much of south Wales, this has proved to be a common species in many areas. It becomes genuinely rare in northern England and ultimately becomes restricted to coastal sites. In Scotland it is known from a few isolated

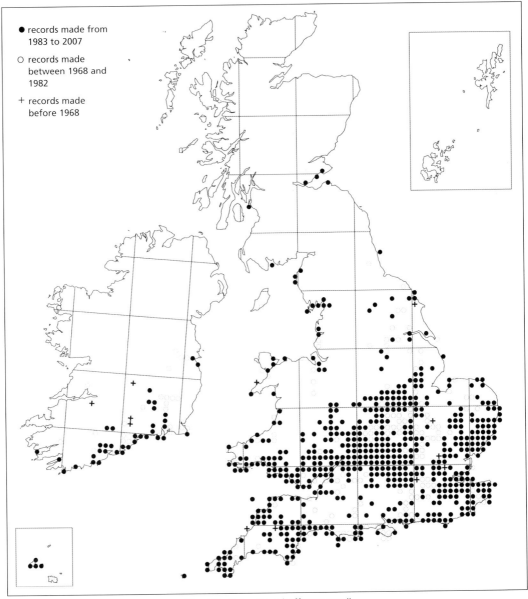

Map 35. *Platyarthrus hoffmannseggii.*

Figure 22. *Platyarthrus hoffmannseggii.* © Paul Richards.

coastal populations north to Ayrshire and Fife. In Ireland it mainly occurs in the south-east, but populations are known as far afield as Counties Kerry and Dublin.

Habitat: This is a myrmecophilous woodlouse and should be sought in the nests of ants, especially *Lasius flavus* or *L. niger* (Hames, 1987; Cawley, 2001). It shows a strong preference for calcareous soils, but this may be because these tend to warm up quickly, rather than because of their lime content. In the southern parts of Britain and Ireland it may be found wherever suitable species of ants occur, including gardens, churchyards, road verges and semi-natural grasslands. In Gloucestershire, *P. hoffmannseggii* has been found associated with arboreal ant nests (*L. flavus* and *L. brunneus*) inside hollow orchard trees (Alexander, 2008). On very rare occasions individuals can be found away from ants' nests.

Microsites: Lifting stones or paving slabs in sunny locations is the easiest way to find an ants' nest. Being white in colour *P. hoffmannseggii* is usually conspicuous against the dark background of the ants' tunnels and galleries. In graveyards it can often be found under flower vases (Daws, 1994a). Unlike most soil-dwelling woodlice, it becomes very elusive throughout the winter months, possibly as a result of moving deep underground until the host ants become active again in the spring.

Associated species: In addition to the common *L. niger* and *L. flavus*, *P. hoffmannseggii* can be found associated with a wide range of other ant species. Hames (1987) records *Formica cunicularia*, *F. lemani*, *F. rufa*, *Lasius alienus*, *L. brunneus*, *L. fuliginosus*, *L. umbratus*, *Myrmica rubra*, *M. ruginodis*,

M. sabuleti, *M. scabrinodis* and *Tetramorium caespitum*. Alexander (2008) reports *L. brunneus* as a host in Gloucester, but also notes that the records given by Hames (1987) are beyond the British range of this southern ant and are therefore erroneous. Robinson (2001) adds *Formica fusca*. In Ireland Cawley (2001) found *P. hoffmannseggii* in nests of *L. niger*, *L. flavus*, *M. ruginodis* and *M. scabrinodis*, but not in those of *T. caespitum* or *Formica* spp. although these latter ants were present at some of the recorded localities.

Other notes: The presence of *P. hoffmannseggii* in arboreal ants' nests (Alexander, 2008), suggests that in southern England, at least, the presence of ants is more critical than the actual habitat that the ants occupy. However, at the northern limits of its range its distribution is not related to that of potential ant hosts. In Cumbria and Lancashire *P. hoffmannseggii* is restricted to coastal limestone areas, despite the widespread occurrence of *Lasius flavus* and *L. niger* on hills up to 311m asl many kilometres inland (Robinson, 2001). In fact both *L. niger* and *L. flavus* are widespread throughout northern England and Scotland (Barrett, 1977). As suggested by Hopkin (1987a), the northern limits of its range appear to be defined by the presence of calcareous soils and high levels of summer insolation.

The precise relationship between *P. hoffmannseggii* and its ant hosts is not fully understood. Although probably mutually beneficial, Williams and Franks (1985) consider that the woodlouse receives the main advantage in the form of constant food supply (general nest detritus and ant faecal pellets have been suggested as a possible sources of food) and protection from predators. It is not known how *P. hoffmannseggii* disperses from nest to nest. However, it is able to recognise its own nest and specimens transferred into a different nest show higher mortality rates (Williams & Franks, 1985). Ants normally tolerate *P. hoffmannseggii*, but if ants become aggressive it has been observed to clamp down and to exude repellent chemicals from its uropods (Gorvett & Taylor, 1960).

Worldwide distribution: This is a common and widespread woodlouse throughout much of Europe, North Africa and Asia Minor and has been introduced to North America (Schmalfuss, 2004).

Oniscus asellus Linnaeus, 1758

Distinctive features: This large grey-brown woodlouse reaches 18mm. Often the epimera are expanded at the sides giving specimens a rather broad, flattened appearance. The dorsal surface is smooth in adults, but juveniles are very rough and could be mistaken for *Porcellio scaber*. However, the three flagella segments and lack of pleopodal lungs are characteristic of O. *asellus*.

Distribution: Having been recorded from all vice-counties in England, Wales, Scotland and Ireland, *O. asellus* is our most widely recorded woodlouse. This species (along with *P. scaber*) represents the archetypal 'woodlouse' familiar to the general public. Unlike most other species, it remains equally abundant even in the most northerly counties of Scotland (Harding & Sutton, 1985) and, in time, it will probably be discovered in every 10km square. Since this is an easily found species, the map essentially indicates areas where woodlouse

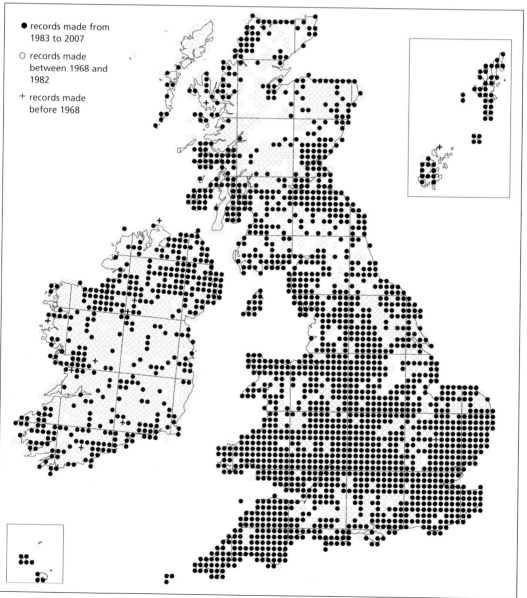

● records made from
   1983 to 2007

○ records made
   between 1968 and
   1982

+ records made
   before 1968

Map 36. *Oniscus asellus.*

Figure 23. *Oniscus asellus.* © Paul Richards.

records have been made. The map should be compared to that of *Trichoniscus pygmaeus*, which is probably just as common in many areas, but is elusive and rarely found unless specifically searched for.

Habitat: From seashore to mountain summit this is a ubiquitous species, but it is particularly abundant in woodland, scrub, hedge banks, damp meadows and on coastal cliffs of rock or clay. It readily invades synanthropic sites such as waste ground, churchyards, gardens, parks and disused quarries. In many southern English counties, such as Oxfordshire (Gregory, 2001), it can be extremely abundant in urban areas, greatly outnumbering *P. scaber*. In other areas, such as Cardiganshire (Chater, 1986a) and around Sheffield (Richards, 1995) it tends to become more abundant in woodland and rural sites.

It is clearly well adapted to northern latitudes and high altitudes and is extremely tolerant of acidic conditions. It is therefore able to successfully penetrate upland moorland, often found on damp ledges or beside watercourses. In Ireland it even inhabits acid blanket bog to the exclusion of other woodlice (Doogue & Harding, 1982). It avoids the driest localities and is rarely found in short turf grassland or open heathland and can be scarce in supralittoral habitats or on sand dunes (where *P. scaber* may be abundant).

Microsites: Although there is considerable overlap, *O. asellus* and *P. scaber* tend to occupy slightly different microsites. *O. asellus* usually occurs close to ground level and large numbers may be found under dead wood or stones. However, it can take shelter in damp crevices of all sorts, including under flowerpots, amongst leaf litter, within compost heaps, amongst rubbish, inside damp bathrooms and kitchens, within rotting window sills and near the base of walls. Inside glasshouses it can be a minor pest of seedlings.

Worldwide distribution: With a broad Atlantic distribution, *O. asellus* is one of the most abundant species in western Europe, but becomes increasing scarce and synanthropic in the north (where it reaches southern Finland) and the east (to Romania). It has been widely introduced into the Americas (Schmalfuss, 2004).

Oniscus asellus ssp. *occidentalis* Bilton, 1994

Taxonomic note: The interpretation of the distribution and ecology of O. *asellus* in south-west Britain is complicated by the occurrence of two morphologically distinct taxa. O. *asellus asellus* Linnaeus is the typical form occurring ubiquitously throughout Britain and Ireland. O. *asellus occidentalis* Bilton is a south-western form first recognised by David Bilton (1990). Considering the distinctive form of the male first endopod and first pereopod this was initially thought to be an undescribed species. However, where the two taxa meet morphological intermediates were found and consequently the two forms were described by Bilton (1994) as subspecies.

Subsequent genetic studies have indicated that O. *a. asellus* and O. *a. occidentalis* represent genetically distinct taxa of ancient genetic divergence, probably as a consequence of fragmentation and isolation of species refugia

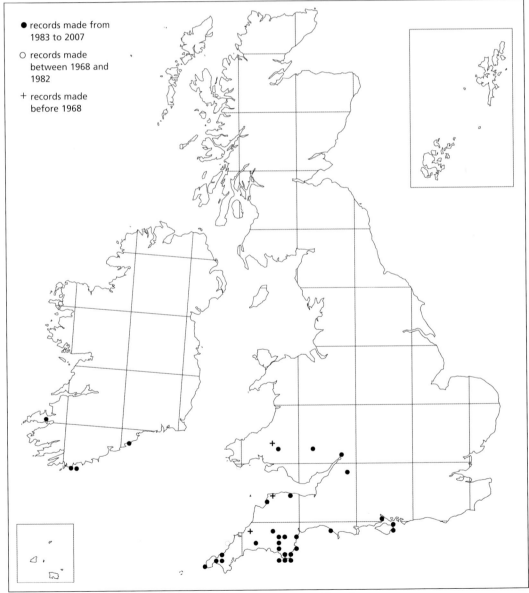

● records made from
   1983 to 2007

○ records made
   between 1968 and
   1982

+ records made
   before 1968

Map 37. *Oniscus asellus* ssp. *occidentalis*.

Figure 24. *Oniscus asellus* ssp. *occidentalis*. © Paul Richards.

during the Pleistocene glacial cycles (Bilton, Goode & Mallet, 1999). *O. a. occidentalis* is believed to be a relict form indigenous to the British Isles. *O. a. asellus* is a competitive form strongly favoured by human activity.

Morphologically intermediate populations were shown to be of hybrid origin, but, atypically for an invertebrate, were analogous to a hybrid swarm as typically seen in plants. Without human influences the two forms may have rarely met and interbred (Bilton, 1994). However, habitat fragmentation has allowed *O. a. asellus* to colonise areas previously dominated by its less competitive sibling *O. a. occidentalis*, which is slowly being hybridised into extinction. Consequently, *O. a. occidentalis* could be considered to be our most vulnerable woodlouse taxon.

Distinctive features: *O. a. occidentalis* is smaller (to 9mm) than typical *O. a. asellus*. The body tends to be more arched in cross section (more akin to *Armadillidium* spp.) and more brightly coloured. However, colour and body shape are not reliable, particularly with intermediate forms, and identifications should be based upon male specimens. The most reliable feature is the presence of a prominent symmetrically forked tip to the male first endopod (Bilton, 1994), readily seen with a hand lens. In intermediate forms this is asymmetric, with one spike smaller, even reduced to a small bump.

Distribution: Pure *O. a. occidentalis* has a generally south-western distribution from West Cornwall to the Isle of Wight and north to south Wales. A few populations have been discovered on the south coast of Ireland. Intermediate forms occur within and beyond this range eastwards to Kent and north to the Isle of Man and Yorkshire.

Habitat: The two taxa differ in their habitat preferences. In south-western areas *O. a. occidentalis* favours semi-natural damp woodland, wetland and rank grassland, whereas *O. a. asellus* readily invades drier places, including farmyards, towns and other synanthropic sites (Bilton, 1994). Intermediate forms occupy transient habitats such as road verges and recent woodland plantation. Further north and east *O. a. occidentalis* becomes increasingly restricted to wet habitats. Ultimately, isolated populations of intermediate forms persist in fragments of semi-natural wetland habitat entirely surrounded by populations of typical *O. a. asellus* (e.g. Lye Valley Fen SSSI in Oxford city (Bilton, 1994)).

Worldwide distribution: *O. a. occidentalis* is mainly known from south-western Britain, but populations have been discovered in the Channel Islands, north western France and an outlying population in the Pyrenees. Intermediate forms occur more widely throughout western Europe (Bilton, 1994).

Armadillidium album Dollfus, 1887

Distinctive features: This pill-woodlouse reaches about 6mm in length. Although easily overlooked as a pale A. *vulgare*, A. *album* has a characteristic stance when disturbed (though it can roll into a loose ball). Its body remains arched allowing a few pairs of protruding legs to grip the underlying substrate. The body surface is covered in minute spines.

Distribution: This uncommon species occurs sporadically around the British coastline, from the Isles of Scilly as far north as Luce Bay in south-western Scotland and Berwick-on-Tweed in north-eastern England. In Ireland it occurs at several sites on the east coast.

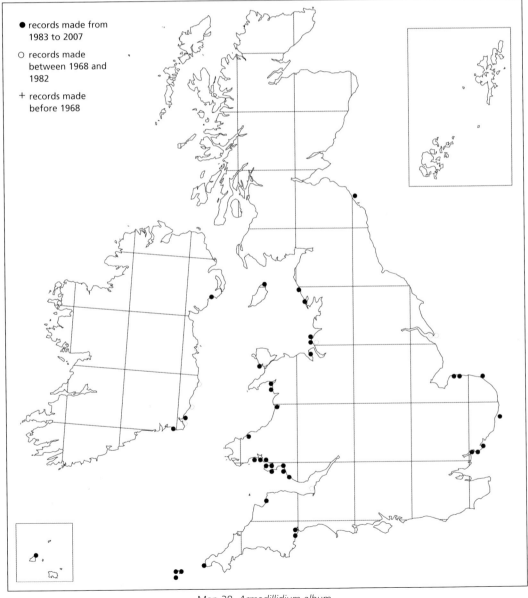

● records made from
  1983 to 2007

○ records made
  between 1968 and
  1982

+ records made
  before 1968

Map 38. *Armadillidium album*.

Figure 25. *Armadillidium album*. © Dick Jones.

Habitat: A. *album* seems to have quite specific habitat requirements. It is strongly associated with undisturbed sand-dune systems, rarely salt marsh, where the affects of human activities and/or strong winds and tides are minimal (Harding & Sutton, 1985). It also seems to favour a narrow range of sand grain size. Consequently, it has not been recorded from many apparently suitable localities. Along the coastline of southern Ireland Cawley (2001) reports little suitable habitat. Although unable to find additional populations, he suggests that new sites may await discovery in Co. Kerry. Although typically associated with storm strandline debris, on occasions it can be found in sandy grassland on the seaward side of fore dunes.

Microsites: Usually, it is found clinging to the underside of driftwood or hiding within crevices, but it may also be found at depths of 20-30cm into underlying sand. Although normally found in small numbers it can sometimes be abundant where large quantities of buried storm strandline debris have accumulated (Chater, 1986b).

Associated species: It is usually associated with *Porcellio scaber* and A. *vulgare*, or occasionally with other species.

Other notes: Considering its penchant for undisturbed semi-natural sand dune systems A. *album* should be considered potentially at risk throughout its British and Irish range. At some sites it is possibly threatened by heavy tourist pressures, excessive tidying (removal) of strandline material and commercial sand extraction. However, at specific sites, such as Ynys-las Dunes NNR, Cardiganshire, it can be locally numerous and Chater (1986a) considered it unlikely to be threatened by heavy tourist pressure which tends to affect the more superficial features of the foreshore.

Worldwide distribution: A. *album* is widely dispersed around the European coastline from the Netherlands in the north and along the Atlantic and Mediterranean coasts as far east as Greece (Schmalfuss, 2004). The populations at Luce Bay and Berwick-on-Tweed are the most northerly known localities in the world.

Armadillidium depressum Brandt, 1833

Distinctive features: This is a large pill-woodlouse reaching 20mm. It is reminiscent of the common *A. vulgare*, but the pleon is more flattened and splayed outwards. When rolled into a ball it characteristically leaves a small gap.

Distribution: With a distinct south-western bias, *A. depressum* can be locally abundant wherever suitable calcareous substrates occur in Devon, Dorset, Gloucestershire, Glamorgan and Pembrokeshire. Since the publication of Harding and Sutton (1985) its known range has extended considerably eastwards, with strong populations discovered in Oxfordshire and Sussex. It has probably been widely dispersed by human activities and Sutton and Harding (1989) suggest

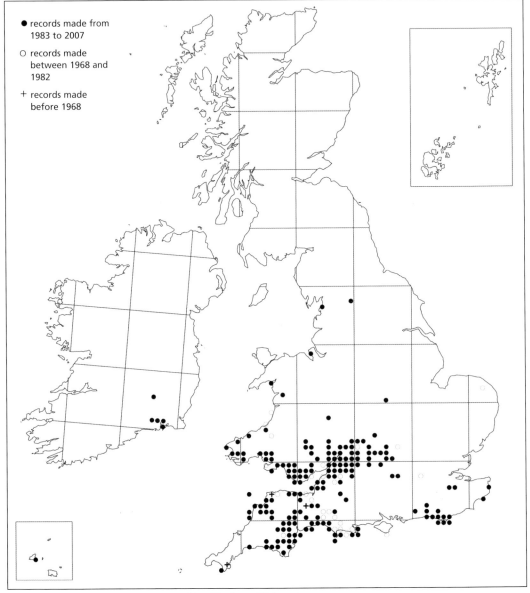

● records made from
  1983 to 2007

○ records made
  between 1968 and
  1982

+ records made
  before 1968

Map 39. *Armadillidium depressum*.

Figure 26. *Armadillidium depressum*. © Dick Jones.

that its patchy distribution may be the result of chance introductions. It may be going through a progressive expansion of range.

The isolated colonies in Leicester, Norfolk, the Wirral (Cheshire), Lancaster and Yorkshire are undoubtedly introductions and it will be interesting to see if the species becomes more widespread in these areas. In 2008 a population was discovered in Cambridgeshire (TL09, not mapped). Recently it has been discovered in Counties Waterford, Wexford and Kilkenny in south-eastern Ireland where it may have been present for many years prior to its discovery (Cawley, 1997, 2001). Waterford City once had strong shipping links with Newport (Monmouthshire) on the Severn Estuary and this is a potential source of introduction (O'Meara, 2002).

Habitat: Throughout its range *A. depressum* is heavily synanthropic, in both rural and urban situations. It favours dry, relatively exposed places and typically occurs on dry limestone walls or loosely mortared walls. It can often be found in gardens and churchyards and can inhabit damp houses, where it may appear in large numbers at night. It also occurs where limestone rocks are exposed, such as in railway cuttings and disused quarries. At one site in Ireland it occurs on a low sea cliff under ivy *Hedera helix* and among stones at the high water mark (Cawley, 2001).

Microsites: During the day it hides away in deep crevices and can be difficult to locate. On walls individuals may be located under loose flakes of stone or mortar, or beneath large capping stones. At night it may be seen in good numbers crawling over walls. In gardens it can be found in rockeries and under paving slabs, especially where these lie on free draining calcareous soil. In quarries and cuttings it occurs amongst limestone rubble and beneath stones. It is a strong burrower and during drought and cold weather it retreats deep into walls and underground crevices, becoming very elusive.

Associated species: Typically, it is associated with *A. vulgare*, *Porcellio scaber* and *P. spinicornis*.

Worldwide distribution: *A. depressum* also occurs in Italy, southern France, western France and Belgium (Schmalfuss, 2004).

Armadillidium nasatum Budde-Lund, 1885
Armadillidium speyeri Jackson, 1923

Distinctive features: This pill-woodlouse reaches 12mm and is reminiscent of *A. vulgare*. The dark grey body is typically marked with pale longitudinal stripes that give a semi-translucent appearance and the narrow protruding scutellum (snout) is distinctive. It forms an imperfect ball when disturbed leaving its antennae protruding.

Distribution: Either side of the Severn Estuary, in parts of Devon and south Wales, and across much of south-eastern England *A. nasatum* can be locally common. Across central, eastern and northern England it occurs sporadically as far north as south Lancashire, with an isolated glasshouse record

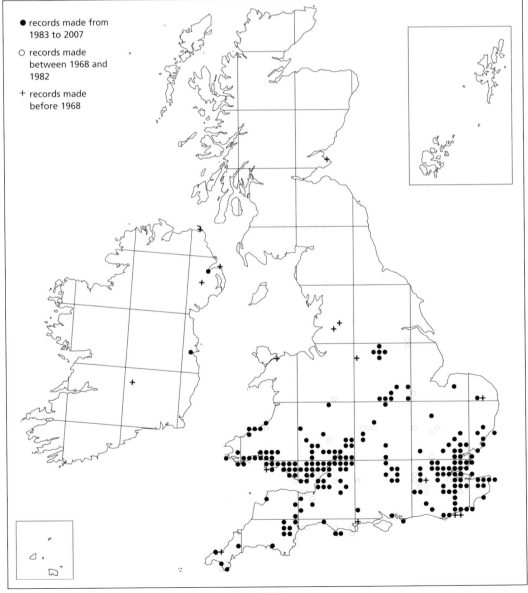

● records made from
   1983 to 2007

○ records made
   between 1968 and
   1982

+ records made
   before 1968

Map 40. *Armadillidium nasatum*.

Figure 27. *Armadillidium nasatum*. © Paul Richards.

in Fife, Scotland. There are few Irish records and just one additional locality discovered since the publication of Harding and Sutton (1985).

Habitat: This species is characteristic of dry, sparsely vegetated habitats, both semi-natural and synanthropic, that are subject to high levels of insolation. Although it favours hot summers it also tolerates cold winters (Hopkin, 1987a). It is usually associated with calcareous substrates. In south-western England and south Wales A. *nasatum* is often associated with coastal grassland, including vegetated 'soft cliffs', dune grassland, limestone screes and open sunny hedge banks. Throughout its range it also occurs in synanthropic habitats, such as disused limestone quarries, railway lines, industrial waste ground, disused limekilns and garden centres. Towards the north of its range it becomes increasingly confined to glasshouses, especially those of botanic gardens and plant nurseries.

Microsites: Individuals tend to aggregate and once a suitable microsite is discovered it is often numerous. It is typically found under rocks, stones, pieces of wood and other debris or amongst rubble.

Other notes: Human activities, including transportation of limestone rock (for buildings, limekilns, etc), aggregates (for road and railway construction) and potted plants (gardens, glasshouses, etc) have undoubtedly spread A. *nasatum* far and wide. Doogue and Harding (1982) consider it to be an introduction into Ireland, where it has a clear association with synanthropic sites, such as glasshouses, conservatories, gardens and parks.

Worldwide distribution: Vandel (1962) considered A. *nasatum* to have originated in the Mediterranean region and to have subsequently spread throughout western Europe. It is introduced and synanthropic in northern and eastern Europe and has been introduced into North America (Schmalfuss, 2004).

Armadillidium pictum Brandt, 1833

Distinctive features: This medium pill-woodlouse, reaching 9mm in length, has the same attractive mottling as *A. pulchellum*. Confusion with both *A. pulchellum* and *A. vulgare* has occurred frequently and Gregory & Richards (2008) highlight the differences between the two species and errors in published keys.

Distribution: A. *pictum* is a rare woodlouse with a marked north-western distribution in Britain. It is listed in the British Red Data Book (Bratton, 1991), but several additional localities have been discovered since the distribution maps of Harding and Sutton (1985) and Richardson (1989) were published. There is a cluster of records extending from the English Lake District, along the

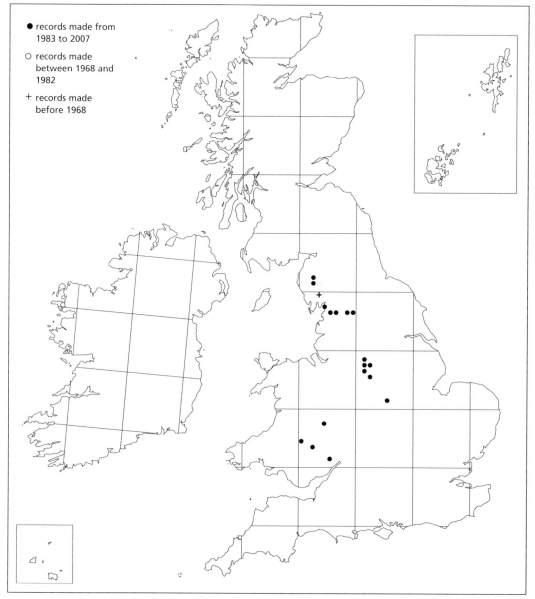

Map 41. *Armadillidium pictum.*

Figure 28. *Armadillidium pictum.* © Paul Richards.

Pennines of Yorkshire and Derbyshire and into Leicestershire (Charnwood Forest, Daws, 1996). A second cluster is centred on the Welsh/English border counties of Breconshire, Radnorshire (Chater, 1988), Gloucestershire (Alexander, 1995) and Herefordshire (Gregory, 2008). This elusive species was discovered in the well-worked county of Derbyshire as recently as 1998 (Richards & Thomas, 1998), where it has subsequently proved to be widespread, but uncommon (Richards, 2004). A. *pictum* may have been overlooked in other areas within its range and it may occur further afield, perhaps in south-western England or southern Scotland.

Habitat: Unlike most other rare British woodlice, A. *pictum* is exclusively associated with semi-natural habitats, typically in hilly areas. Many localities are ancient deciduous woodland, but it also inhabits scrubby, rough and/or shady grassland, including grikes in limestone pavement. The presence of suitable rocky terrain, such as talus slopes with accumulations of scree, rocks or boulders, seems to be a common feature of most sites. Harding (2006) describes its occurrence within tufa screes at Tarren yr Esgob. It shows no obvious preference to rock type, being found in both calcareous and acidic localities. However, some non-calcareous strata, such as those of the Borrowdale volcanic series, produce base-rich screes as a result of constant land-slippage and rock-fall (K.N.A. Alexander, personal communication). A. *pictum* seems to be able to locate and exploit such base-rich features within an otherwise non-calcareous landscape (P.T. Harding, personal communication).

Microsites: It is generally found at ground level beneath thick moss carpets, under stones, amongst scree or within red-rotted timber. It will climb vegetation, especially in wet conditions, and can be beaten from shrubs such as juniper *Juniperus communis*. In deciduous woodland it may be found inhabiting loose bark, rot holes and other 'dead wood' habitat on trees up to two metres above ground level (Chater, 1988; Richards & Thomas, 1998). Unlike A. *pulchellum*, it does not seem to occur in short-turf swards subject to high insolation.

A. pictum is able to occupy a wide variety of microsites, both between different localities and at any one given site. Despite this, its occurrence at many sites appears to be patchy, with several specimens being found within a small area, whilst it is apparently absent from suitable habitat nearby (e.g. Gregory, 2008). On occasions, especially in spring, it may be found in superficial habitats only to become very elusive on subsequent visits, presumably as a result of retreating deep into crevices. Certainly, talus slopes will allow the species to readily disperse underground.

Associated species: Few woodlice, other than *Porcellio scaber* and *Oniscus asellus*, tolerate non-calcareous conditions, but on limestone *A. pictum* is associated with a wider variety of species, including *Philoscia muscorum* and *Trichoniscus pusillus* agg., and, where their respective distributions overlap, *A. vulgare* (Gregory, 2008).

Other notes: Bratton (1991) considered possible threats to this rare species to be the removal of rock (e.g. limestone pavement) for ornamental use and increased tourist pressure to the 'beauty spots' where it occurs. Although genuinely rare, because of its elusive nature, it is also likely to be relatively under-recorded. Thus, it is not considered to be under immediate threat.

Worldwide distribution: This is a species of north western Europe, occurring from southern Sweden and Denmark, through Germany, the Netherlands, Belgium, Britain, France and Switzerland (Oliver & Meecham, 1993).

Armadillidium pulchellum (Zencker, 1798)

Distinctive features: A. *pulchellum* is a small attractively patterned pill-woodlouse reaching 5mm in length. It is reminiscent of a well-patterned A. *vulgare* juvenile or a mottled immature pill-millipede *Glomeris marginata*. However, confusion is most likely with A. *pictum*, which is almost identical in appearance and may occupy similar microsites (Gregory & Richards, 2008).

Distribution: The striking north-western distribution extends from Devon and Cornwall in the southwest, through Wales and northern England as far north as the Solway coast of south-western Scotland. An outlying population has been recently discovered at Arbroath, Angus on the Scottish east coast. A few isolated populations are known in south-eastern England. In Ireland, it

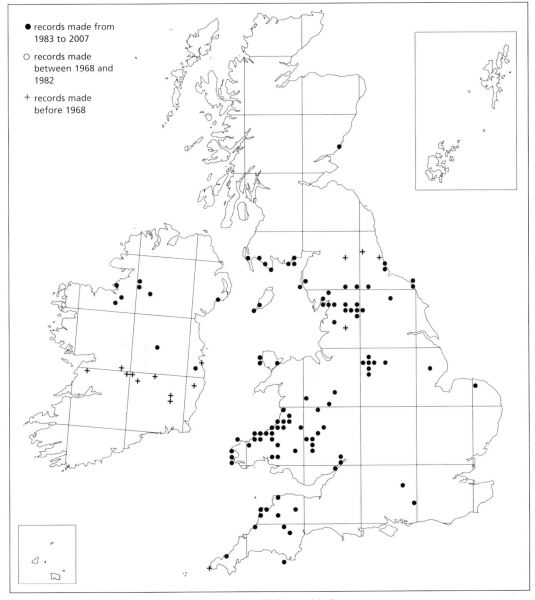

- ● records made from 1983 to 2007
- O records made between 1968 and 1982
- + records made before 1968

Map 42. *Armadillidium pulchellum*.

101

Figure 29. *Armadillidium pulchellum*. © Paul Richards.

is widespread across the midlands (where few records have been confirmed since the 1980s) and in the northwest.

Habitat: This is primarily a woodlouse of rural semi-natural habitats. Although it may be found on acidic substrates it generally favours calcareous soils. At some sites, A. *pulchellum* is able to find and survive in tiny isolated pockets with calcareous features within an otherwise acidic area (P.T. Harding, personal communication). In Britain and Ireland it is mainly associated with open grassy habitats, but there seems to be some degree of regional variation. These can be broadly split into five distribution patterns; coastal, Irish Midlands, upland England, Wales and Devon and south-eastern England, as described below.

Throughout its range A. *pulchellum* is widely associated with coastal grasslands and becomes increasingly confined to this habitat in more northern regions. Inland in Ireland, it typically inhabits calcareous grasslands associated with glacial deposits in the Irish midlands (Doogue & Harding, 1982). In upland Britain, such as the Carboniferous limestones of northern England, it typically inhabits short turf grasslands on south facing slopes, sparsely vegetated screes and limestone pavement. Further south in Devon and Wales it occupies a wider variety of habitats, including rough grassland, stands of heather *Calluna vulgaris* or bracken *Pteridium aquilinium*, shady banks, vegetated stone walls and, in east Wales, Chater (1989) also records oak *Quercus* sp. woodland. Many of these latter habitats are akin to those favoured by A. *pictum*. The few records in southern and eastern England are associated with lowland heath, typically where planted or self-sown conifers are present. These are probably relict populations and A. *pulchellum* is likely to be present on other English lowland heaths. To a certain extent these distribution patterns reflect an increasing requirement for higher levels of insolation in northern areas.

Microsites: Often *A. pulchellum* forms small colonies restricted to a few square metres where the microclimate is damper than surrounding areas. Typically it occurs under stones or mats of plants (such as stonecrop *Sedum* sp. and thyme *Thymus* sp.) or among leaf litter, grasses, lichens and mosses. It may be associated with ants, including *Lasius flavus*, *Myrmica* sp. and wood ants *Formica* sp. (Morgan, 1994). In wet conditions it will climb vegetation, such as bracken and heather, and may be collected by sweep net. On lowland heathland specimens have been collected beneath the bark of a dead conifer (Hopkin, 1986), under dead wood and among litter below heather plants (Alexander, 2000; Telfer, 2007) and in a wood ant *Formica* sp. nest (A. Fowles, personal communication).

Other notes: Where their ranges overlap *A. pulchellum* and its congener *A. vulgare* do not usually co-exist. *A. pulchellum* tolerates high altitude and non-calcareous substrates, *A. vulgare* does not. However, in coastal situations or on English lowland heaths both species may be found. In contrast, there is considerable overlap in the distribution and habitat preferences of *A. pulchellum* and *A. pictum*. Generally *A. pulchellum* tolerates higher levels of insolation, but observations in Wales indicate that this distinction is not reliable (Chater, 1989; J.F. Harper, personal communication). Gregory and Richards (2008) stress that care is needed with the identification of these three species.

Worldwide distribution: According to Schmalfuss (2004), this species occurs throughout Europe, except the Mediterranean and south-eastern regions. However, Britain and Ireland hold the greatest proportion of the known sites in Europe (Harding & Sutton, 1985).

Armadillidium vulgare (Latreille, 1804)

Distinctive features: This large pill-woodlouse, reaching 18mm in length, is able to form a perfect sphere with all appendages concealed. Pigmentation is variable, but typically uniform slate grey. Some populations may show ornate mottling, but never as well developed as seen in A. *pulchellum* or A. *pictum*.

Distribution: The only common and widespread pill-woodlouse occurring in Britain and Ireland, A. *vulgare* is locally abundant in south-eastern England. It begins to become less numerous in northern Oxfordshire (Gregory & Campbell, 1995), but it remains widespread and frequent south of a line from south-east Yorkshire to the Severn Estuary. North of this line it becomes

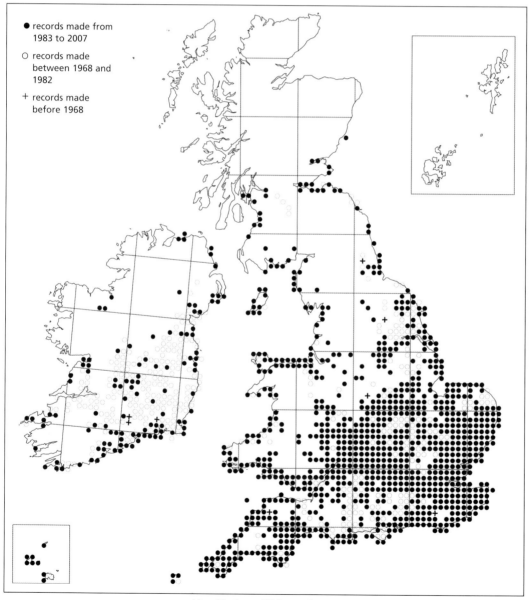

Map 43. *Armadillidium vulgare.*

Figure 30. *Armadillidium vulgare.* © Paul Richards.

increasingly local and predominantly coastal. It occurs north to the Clyde (Glasgow) in the west (Futter, 1998; Collis, 2006, 2007) and to Kincardineshire in the east. In Ireland it is widespread over much of the south-east as far north as Co. Down and Co. Clare. The Irish distribution closely follows that of limestone outcrops and it is scarce and mainly coastal in the south-west where the rocks are predominantly non-calcareous. Outlying populations occur as far north as the coast of Counties Antrim and Londonderry.

Habitat: A. *vulgare* is restricted to calcareous soils, unless the effects of low pH are ameliorated by maritime or human influences. It also favours sites with high insolation and is often active by day, even in full sun. It is intolerant of high altitudes and in the English Midlands it is rarely found above 100m asl (G.M. Collis, personal communication). On the Carboniferous limestones of northern England it is entirely replaced by its congener A. *pulchellum*, a species well adapted to these upland areas. In coastal habitats the two species may co-exist.

Over much of south-eastern England it is common in most habitats, including gardens, waste ground, arable margins, quarries, road verges, rank grassland, short turf species-rich swards and sand dunes. In the Sheffield area, near the edge of its inland range, A. *vulgare* occurs in semi-natural grasslands on Permian magnesian limestone in the east and in synanthropic habitats such as collieries, railway lines, canals and disused tips. In contrast, A. *pulchellum* occurs exclusively on the Carboniferous limestones in the west of the survey area (Richards, 1995).

At its northern limits in Britain and Ireland A. *vulgare* typically occurs at well-drained localities with high insolation, such as maritime grasslands or synanthropic localities, including railway lines, near the coast. The few inland Scottish records are mostly from gardens or glasshouses (Harding & Sutton, 1985). It is probable that railways, which often follow the coastline in Scotland, provide a corridor for dispersal, most likely following introduction via garden rubbish.

In Sligo and Mayo, Ireland, outlying populations occur along the disused Sligo-Limerick railway where the species was probably introduced with ballast during construction (Cawley, 1996).

Microsites: It is most readily found under stones and dead wood, but also amongst grass litter, within tussocks, etc. It is usually found with other large species, particularly *Philoscia muscorum* or *Porcellio scaber* and on the coast *Porcellionides cingendus*, *Cylisticus convexus*, *A. pulchellum* or, on sandy beaches, *A. album*.

Worldwide distribution: Originating in the Mediterranean region, *A. vulgare* occurs throughout most of Europe with the exception of northern and upland areas. It has been introduced to all parts of the world (Schmalfuss, 2004).

Eluma caelatum (Miers, 1877)
Eluma purpurascens Budde-Lund, 1885

Distinctive features: This large pill-woodlouse, reaching up to 15mm in length, is virtually identical to *Armadillidium vulgare* in size and shape and readily forms a perfect sphere upon disturbance. It is easily distinguished by possessing eyes composed of single prominent ocelli. It is typically purplish-brown in colour and the body appears slightly hairy.

Distribution: First discovered in Ireland on the cliffs of Howth, Co. Dublin in 1908, *E. caelatum* has proved to be common along much of the Co. Dublin coastline and is known from a few sites inland (Harding & Sutton, 1985). For many decades it was unknown in Britain, until its discovery at

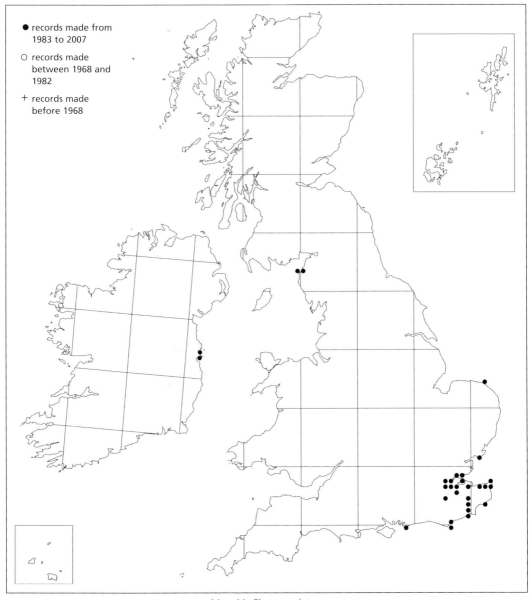

● records made from
1983 to 2007

○ records made
between 1968 and
1982

+ records made
before 1968

Map 44. *Eluma caelatum.*

Figure 31. *Eluma caelatum.* © Theodoor Heijerman.

Overstrand, East Norfolk in 1975 (Harding, 1976b) and five years later near Whitstable, East Kent. It is now apparent that *E. caelatum* has a wide distribution across the extreme south-east of England, both on the coast and inland. In 1995 two outlying populations were found on the Cumberland coast, northern England. These represent the most northerly populations for this species in the world.

Habitat: Semi-natural localities include eroding sparsely vegetated 'soft rock' cliffs and in Ireland also sand dunes and salt marsh drift line, where *E. caelatum* occurs in the splash zone and above (Doogue & Harding, 1982). It readily takes to synanthropic habitats, including sandy waste ground, railway lines and gardens.

Microsites: In coastal situations it typically takes refuge under mat-forming plants, beneath stones and dead wood or among leaf-litter and tussocks. It readily burrows into soil and may be found many centimetres underground. In synanthropic situations it may occur in large numbers under stones and dead wood (including railway sleepers) or among rubbish and other debris.

Associated species: Information on associated species is vague, but *Porcellio scaber*, and in Ireland *Porcellionides cingendus* are both cited. It is of note that at the Cumberland localities *E. caelatum* and *A. vulgare* occupied different microsites (Bilton, 1995).

Other notes: It is probable that this species is a long established introduction in Ireland (Corbet, 1962; M. Cawley, personal communication), if not Britain also. Colonisation of many inland sites seems to have been the result of human activities. In Ireland the few inland localities are railway lines constructed from ballast quarried on the coast (Doogue & Harding, 1982). Additional sites have not been discovered on the south coast of Ireland (M. Cawley, personal communication) which is unexpected if this species had a 'natural' Atlantic distribution. The isolated Cumberland localities, Maryport and Workington, were both major ports and introduction via imported goods is likely (Bilton, 1995). It is probable that further populations will be discovered at other coastal localities near major ports.

Worldwide distribution: The world distribution stretches along the entire Atlantic coast of Europe from the British Isles to North West Africa and includes many offshore islands. It has been introduced into French Guiana (the type locality) and Tasmania (Schmalfuss, 2004).

Cylisticus convexus (De Geer, 1778)

Distinctive features: When provoked C. *convexus* can roll into a flattened ball, but its antennae and long pointed uropods remain protruding. This it quite distinct from the truncated uropods seen in *Armadillidium* spp. and *Eluma caelatum*. This is one of two woodlouse species with five pairs of pleopodal lungs (the other being *Trachelipus rathkii*).

Distribution: This species is widely distributed throughout Britain, but has yet to be recorded from the northern Isles of Orkney and Shetland. It seems to have a patchy distribution, possibly with an eastern bias, but this may be partially due to recorder bias. In Ireland C. *convexus* is widespread

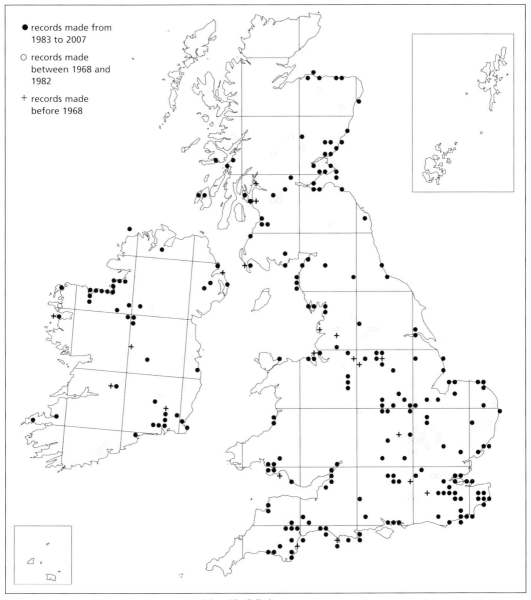

● records made from
   1983 to 2007

○ records made
   between 1968 and
   1982

+ records made
   before 1968

Map 45. *Cylisticus convexus.*

Figure 32. *Cylisticus convexus.* © Paul Richards.

but local (Cawley, 2001), but is considered to be under-recorded in many areas. Cawley (2001) considers it may be a naturalised introduction in Ireland.

Habitat: This is a xerophilous species that tolerates considerable exposure and site disturbance. It is equally at home in sparsely vegetated semi-natural coastal habitats and synanthropic sites inland. *C. convexus* is often found on sparsely vegetated cliff faces and unstable screes, and is often associated with water seepage on 'soft rock' cliffs. It also occurs on shingle beaches, erosion banks and salt marshes, usually where sparse pioneer vegetation has begun to colonise above the strandline. Undoubtedly aided by human activities, it has become widespread in synanthropic situations and inhabits a wide variety of sparsely vegetated synanthropic sites. These include railway sidings and embankments, disused quarries, colliery spoil, farmyards, urban waste ground, rubbish tips, churchyards and gardens. In Ireland the majority of records are from synanthropic sites, such as glasshouses and gardens, often in towns (Cawley, 1996, 2001).

Microsites: It is typically found beneath loose stones, pieces of timber and among other debris, including strandline material. Although sometimes found in large numbers, *C. convexus* can be an elusive woodlouse and at some known sites it has proved difficult to relocate the species (Collis & Collis, 2004).

Associated species: It is typically associated with other 'large species' such as *Philoscia muscorum*, *Porcellio scaber*, *Armadillidium vulgare*, *Oniscus asellus* and also *Trichoniscus pusillus* agg..

Worldwide distribution: This widely distributed species is found throughout Europe and Asia Minor and has been introduced to northern Africa and North and South America (Schmalfuss, 2004).

Porcellio dilatatus Brandt, 1833

Distinctive features: This large species, reaching 15mm, has a pale 'dusty' grey brown coloration. It is relatively broader than *P. scaber*, even when young, and a well grown specimen with the characteristic rounded tip to the telson is quite distinctive. The combination of two flagella segments and two pairs of pleopodal lungs distinguishes it from *Oniscus asellus*.

Distribution: Records of this apparently uncommon species are scattered throughout Britain and Ireland, with isolated records as far north as the Outer Hebrides of Scotland.

Habitat: *P. dilatatus* is almost entirely associated with synanthropic habitats, but mainly in rural situations. It is often found in or around stables, often within well-established manure heaps. Jon Daws (1994a) has

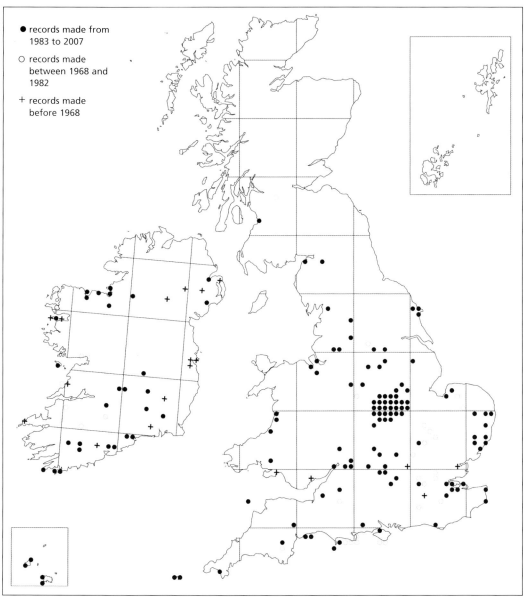

● records made from
   1983 to 2007

○ records made
   between 1968 and
   1982

+ records made
   before 1968

Map 46. *Porcellio dilatatus.*

111

Figure 33. *Porcellio dilatatus.* © Paul Richards.

shown this to be a common and characteristic species of dairy farms in Leicestershire, often associated with *Porcellionides pruinosus*. Subsequently, Cawley (1996, 2001) considered it to be a predictable species of old cowsheds in Ireland, where it is associated with other scarce invertebrates. On rare occasions it can be found on pasture or field margins, presumably introduced via the application of farmyard manure.

P. dilatatus also inhabits churchyards, gardens and allotments, typically within compost heaps and inside glasshouses. Some of the older Irish records are from glasshouses and nurseries (Doogue & Harding, 1982), but Cawley (1996) also records the species from waste ground in Co. Sligo. Harding and Sutton (1985) report an association with ruins. In Cardiganshire, west Wales the species has been recorded from disused limekilns near the coast, probably introduced with imported limestone (Chater, 1986a). A few occurrences are from semi-natural 'soft' cliffs on the coastline of southern England and western Ireland.

Microsites: Specimens of *P. dilatatus* can be encountered under planks, stones and pieces of manure or amongst straw and other debris both inside and outside farm buildings. It can also be found within well-rotted dung heaps and, to a lesser extent, compost heaps. Despite its large size, specimens are usually found with difficulty and often in small numbers.

Other notes: It is likely that this large, but elusive, species is considerably under-recorded. If Leicestershire is typical, then this species could prove to be common on dairy farms across much of Britain and Ireland. It is one of the few woodlice recovered from archaeological excavations. In Leicester city mineralised remains were collected from Medieval rubbish pits (c. AD 1100-1400), along with *O. asellus*, *P. scaber* and *Armadillidium vulgare* (Daws, 1999). There is circumstantial evidence that it has become less common during the 20th Century, possibly reflecting the decline in horses and stables (Harding & Sutton, 1985).

Worldwide distribution: *P. dilatatus* occurs widely across Europe and has been introduced to many other parts of the world (Schmalfuss, 2004).

Porcellio laevis Latreille, 1804

Distinctive features: Another large woodlouse, to 20mm, but unlike other species of *Porcellio* the dorsal surface is smooth. The combination of two flagella segments and two pairs of pleopodal lungs distinguishes it from *Oniscus asellus*.

Distribution: *P. laevis* has been widely recorded across Britain and Ireland, but seems to be most frequent in south-eastern England. There are clusters of records around The Wirral on the Welsh/English border and from the Dublin area in Ireland. Isolated records occur as far afield as Penzance in Cornwall, Glasgow in Scotland and North Kerry in Ireland.

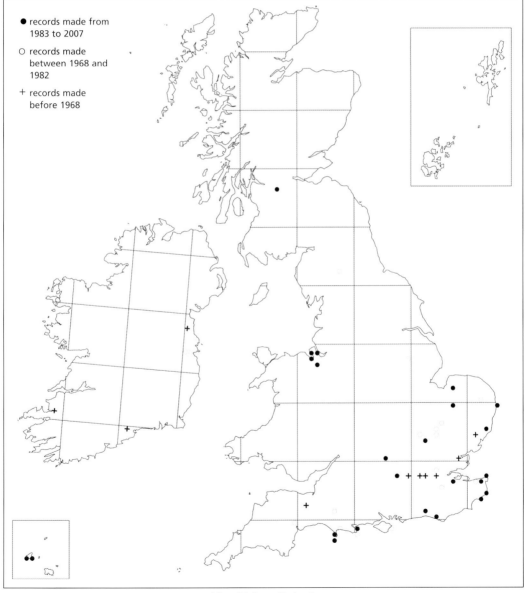

● records made from
1983 to 2007

○ records made
between 1968 and
1982

+ records made
before 1968

Map 47. *Porcellio laevis*.

Figure 34. *Porcellio laevis.* © Dick Jones.

Habitat: This species is principally associated with synanthropic habitats. It often occurs in compost heaps in old well-established gardens, often in suburban locations, but also in major cities. In Oxford city centre it occurs in compost heaps at the Botanical Gardens and in a well-established garden nearby (Gregory & Campbell, 1995). Stables and dairy farms, both inside and outside buildings, are also favoured, but it is much less frequently recorded than *P. dilatatus*.

Microsites: It should be sought within well-rotted compost heaps and dung heaps or under pieces of manure, straw and other debris, wherever some moisture has been retained.

Associated species: Typically, it is associated with *Porcellionides pruinosus* and *Porcellio scaber*. Around the Wirral and North Wales *P. laevis* and *P. dilatatus* co-exist (Daws, 1992a), but interestingly in Leicestershire, where *P. dilatatus* is widespread, *P. laevis* does not occur (Daws, 1994a).

Other notes: A considerable number of records were made in the late 19th and early 20th centuries (Harding & Sutton, 1985), suggesting that this species has declined in recent decades. This is probably due to its penchant for farmyards and stables, which have become fewer and tidier as the motorcar and modern intensive farming practices have taken over. Nonetheless, considering its specialist habitat it is likely to be under-recorded, particularly in areas such as North Wales where Daws (1992a) considers that it could prove to be widespread.

Worldwide distribution: *P. laevis* occurs throughout Europe and North Africa and has been introduced to most other parts of the world (Schmalfuss, 2004).

Porcellio scaber Latreille, 1804

Distinctive features: This is a large woodlouse, reaching 17mm in length, with a heavily tuberculate body. It is typically slatey-grey with the base of the antennae pale orange. Brightly mottled varieties are frequent, but the pattern is never arranged in longitudinal stripes as seen in *P. spinicornis* or *Trachelipus rathkii*.

Distribution: *P. scaber* is our second most frequently recorded woodlouse and has been recorded from all vice-counties in England, Wales, Scotland and Ireland. It is generally ubiquitous and occurs equally in coastal and inland localities. There is no indication of records becoming less frequent in northern counties and, as with *Oniscus asellus*, this species represents the archetypal 'woodlouse' familiar to the general public.

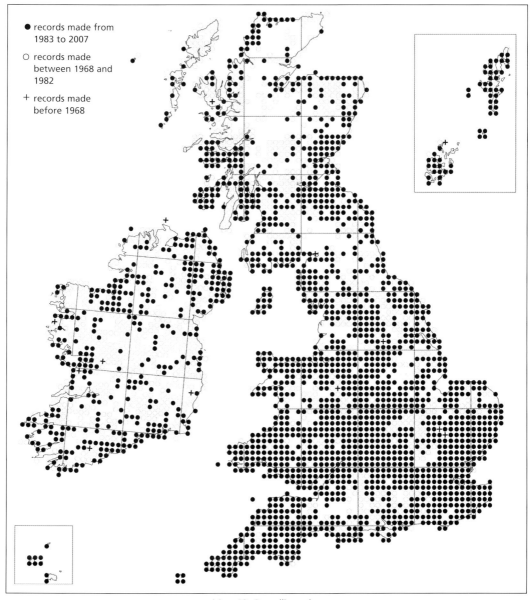

● records made from 1983 to 2007

○ records made between 1968 and 1982

+ records made before 1968

Map 48. *Porcellio scaber.*

Figure 35. *Porcellio scaber.* © Paul Richards.

Habitat: Although there is considerable overlap with O. *asellus* it is apparent that P. *scaber* tends to occupy subtly different, often drier, places. In woodland, grazed grassland and heathland it is particularly abundant. This species also has a liking for coastal habitats, including the supralittoral zone, and may be abundant in sand dunes, maritime grassland, sea cliffs and shingle beaches (habitats in which O. *asellus* is scarce). It also occurs abundantly in gardens, churchyards and waste ground and in many areas, such as Cardiganshire (Chater, 1986a) and the Sheffield area (Richards, 1995), it tends to be more frequent in synanthropic sites. This is the woodlouse most commonly encountered inside houses and it can be a minor pest inside glasshouses. Unlike O. *asellus*, it fails to penetrate far into acidic upland moorland or blanket bogs. However, its association with ruins in upland areas (Chater, 1986a) suggests this is possibly because it is less tolerant of high acidity rather than high altitude.

Microsites: The subtle differences in habitat preferences relative to O. *asellus* are also reflected in microsite selection. As with O. *asellus*, it can be abundant under stones and dead wood, but (unlike O. *asellus*) it is frequently encountered well above ground level. On mortared walls and ruins, especially when covered with ivy *Hedera helix*, it can occur abundantly several metres above ground level. During the summer months it readily ascends trees, especially in woodlands, and may be frequent under loose bark high in the canopy.

Worldwide distribution: P. *scaber* has an Atlantic distribution and is most abundant in western areas. However, it has spread throughout much of Europe as far north as Iceland and northern Scandinavia. It becomes increasingly synanthropic in the east where it reaches the Ukraine (Vandel, 1962) and it has been introduced to many parts of the world (Schmalfuss, 2004).

Porcellio spinicornis Say, 1818
Porcellio pictum Brandt & Ratzeburg, 1833

Distinctive features: This is a large, attractively marked woodlouse up to 12mm in length. The yellow and brown mottling is arranged in a longitudinal pattern and the head tends to be conspicuously dark. This is quite unlike the random mottling seen in *P. scaber*. Confusion is more likely with *Oniscus asellus*, but *Porcellio* spp. are distinguished by having two flagella segments and two pairs of pleopodal lungs.

Distribution: *P. spinicornis* occurs widely throughout Britain and Ireland and can be a locally common species wherever major limestone outcrops occur, such as the Cotswolds and the

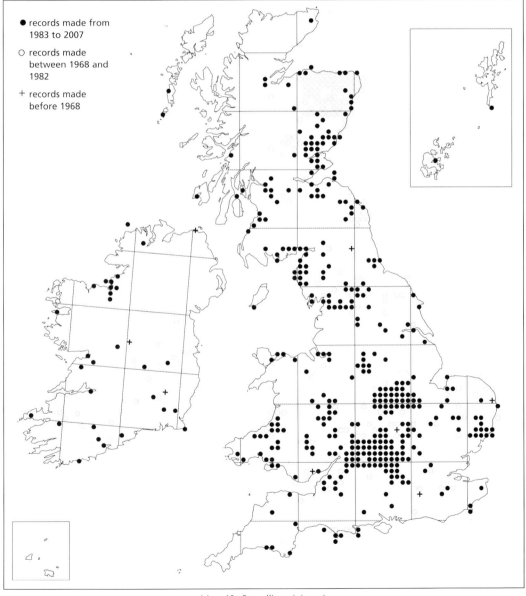

Map 49. *Porcellio spinicornis*.

117

Figure 36. *Porcellio spinicornis*. © Paul Richards

Carboniferous limestones of northern England. Elsewhere, it has proved to be frequent whenever suitable substrates have been specifically searched, whether this is rock outcrops, drystone walls or mortared walls. It has been recorded from all major offshore islands, including the Isle of Man, the Outer Hebrides, the Orkneys, the Shetland Islands, and many of the smaller islands such as Lundy, the Isles of Scilly and Tory (Co. Donegal).

This species is now better known from western regions, including recent records from western Scotland (Collis, 2007; Gregory, 2005) and from Counties Sligo and Leitrim, western Ireland (Cawley, 1996). The north-eastern 'anti-Atlantic' bias described in Harding and Sutton (1985) appears to be largely an artefact of recorder bias. Although *P. spinicornis* remains seriously under-recorded in many areas, it does appear to be genuinely scarce in south-western England (despite being recorded from the Isles of Scilly) and in Pembrokeshire, south-west Wales (Chater, 1986a).

Habitat: P. spinicornis favours dry, relatively exposed, calcareous substrates, such as walls, quarries, cuttings, cliffs and limestone pavement. It is clearly tolerant of cold conditions and occurs in upland areas. Throughout its range it is heavily synanthropic, exclusively so in Ireland (M. Cawley, personal communication), and has probably been widely spread by human activities. It is perhaps most familiar associated with dry limestone walls or loosely mortared walls in either rural or urban situations. It is associated with human habitation and frequently ventures indoors, particularly in western Scotland (G.M. Collis, personal communication) and during inclement weather generally.

It also occurs where limestone rocks are exposed, such as in railway cuttings, disused quarries and in Cardiganshire, west Wales on ruined limekilns and remote ruins (Chater, 1986a). Apparently semi-natural habitats include sparsely vegetated limestone sea-cliffs, upland limestone pavement (where it can be found with *Armadillidium pulchellum*) and limestone screes (Richards, 1995). Atypically, it has been collected in pitfall traps set in coastal shingle in Suffolk (Lee, 2003).

Microsites: On walls individuals may be located under loose flakes of stone or mortar, but lifting capping stones can be productive. In gardens it also occurs in rockeries and under paving slabs lying on free-draining calcareous soil. Where limestone rocks are exposed it can be found amongst rubble and beneath stones. Although it occupies inaccessible crevices by day it may be readily found at night, often in large numbers, crawling over the surfaces of walls or limestone rocks. It can ascend several metres up trees with calcareous bark, such as lime *Tilia* sp. and apple *Malus* sp., where it takes shelter under loose bark.

Associated species: On walls and rocks it is typically associated with *Armadillidium vulgare*, *P. scaber* and, locally, *Armadillidium depressum*. Other frequent associates include *Trichoniscus pusillus*, *Philoscia muscorum* and *O. asellus*.

Worldwide distribution: *P. spinicornis* is a continental species occurring throughout northern and central Europe and has been introduced to North America (Schmalfuss, 2004).

Acaeroplastes melanurus (Budde-Lund, 1885)
Metoponorthus melanurus Budde-Lund, 1885

Distinctive features: Roy Anderson (2007) has observed Irish specimens of *Acaeroplastes melanurus* both in the field and in culture. These are consistently 5 to 6mm in length with a rather narrow, parallel-sided body. Coloration is mottled yellow-brown to grey-brown, but characteristically the head and the pleon are noticeably darker. An irregular surface sculpturing appears dull and slightly rough to the naked eye giving the impression of a mottled juvenile *Porcellio scaber* (though the stepped body outline is more akin to *Philoscia muscorum*).

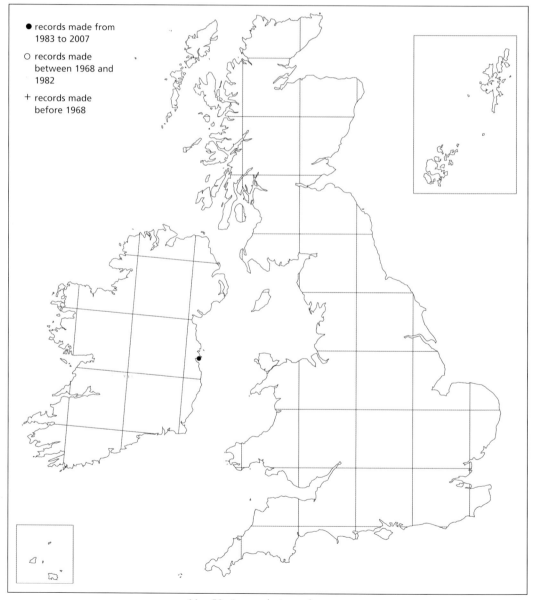

● records made from 1983 to 2007

○ records made between 1968 and 1982

+ records made before 1968

Map 50. *Acaeroplastes melanurus*.

Distribution: *A. melanurus* is only known from a one-kilometre length of the southern cliffs of Howth in Co. Dublin, Ireland. First discovered in October 1909, it was regularly seen for several decades until September 1934 (Doogue & Harding, 1982). Repeated searches in the 1970s failed to re-find the species and Harding and Sutton (1985) considered it extinct.

After an absence of 68 years the species was re-discovered at Howth in April 2002 when a single female was collected from amongst soil underneath stones and vegetation (Wickenberg & Reynolds, 2002). Subsequently, it has been found in good numbers (Anderson, 2007). In November 2006 around 50 specimens were found sheltering under the lichen *Ramalina siliquosa* covering an exposed boulder just below the summit of the south-facing cliffs. Searches in similar microsites nearby were unproductive. In February 2007 a further 10 specimens were found with difficulty in several scattered locations.

Habitat: *A. melanurus* is a xerophilic species, no doubt favoured by the relatively low rainfall and high insolation that occurs at Howth. It is restricted to steep sparsely vegetated slopes where outcrops of lichen-covered rocks occur. Here the sandy/gravely soils are shallow and non-calcareous.

Microsites: Resting sites seem to be dictated by current weather conditions. During inclement weather the species takes refuge amongst lichens covering exposed rocks, often occurring with *P. scaber*. In dry conditions it is believed the species moves down into soil, becoming very elusive.

Other notes: It is clear that a well-established population survives at Howth Head. Although circumstantial, the relatively high numbers found recently suggest it is more plentiful now than during the early 20th Century. Anderson (2007) offers two plausible reasons for the species' apparent demise at Howth. Firstly, a period of unfavourable climatic conditions, due to the exceptionally cold winters of 1947 and 1963 and a period of cool wet summers, caused a marked mid-20th Century decline. Secondly, the ecology of *A. melanurus* was not fully understood and its limited range, cryptic behaviour and weather-dependent niche selection made it an easy species to miss. It is possible that long-term climate change may result in further resurgence of this species. However, the extent of suitable habitat at Howth has contracted significantly since the early 20th Century (Anderson, 2007), mainly due to scrub encroachment. If the habitat continues to diminish then *A. melanurus* may still be threatened with extinction.

Worldwide distribution: *A. melanurus* is essentially a Mediterranean species, occurring in coastal areas from the Spanish/French border to northwest Italy and along the Algerian coast. Outlying populations occur in Croatia, Spain, Azores (introduced) and Ireland (Vandel, 1962; Schmalfuss, 2004).

Porcellionides cingendus (Kinahan, 1857)
Metoponorthus cingendus (Kinahan, 1857)

Distinctive features: This medium sized species, reaching up to 9mm in length, is mottled in various shades of yellow, red or brown. In the field the stepped body outline and fast movement is reminiscent of *Philoscia muscorum* or *Ligidium hypnorum*. *P. cingendus* lacks the distinct dark dorsal stripe seen in *P. muscorum*, whilst *L. hypnorum* tends to be much darker. The antennal flagella of *P. cingendus* are comprised of two segments.

Distribution: *P. cingendus* has a markedly Atlantic distribution across Ireland and Britain. First collected and described new to science from Dublin, it occurs widely throughout the southern

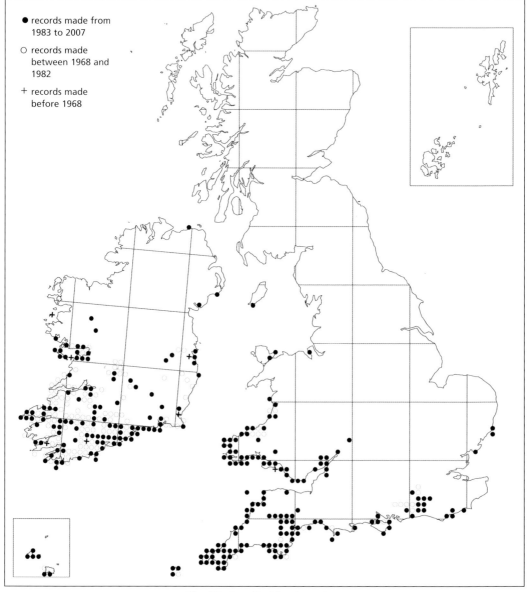

● records made from
1983 to 2007

○ records made
between 1968 and
1982

+ records made
before 1968

Map 51. *Porcellionides cingendus.*

Figure 37. *Porcellionides cingendus.* © Paul Richards.

third of Ireland. In the extreme south-west (Counties Cork and Kerry) it may be the most frequently encountered woodlouse (Doogue & Harding, 1982). It occurs as far north as Co. Mayo on the west coast and Co. Down in the east. In England and Wales it also has a distinctly south-western distribution, ranging from the Isle of Man in the north west, to Orford Ness, East Suffolk in the east. In Britain *P. cingendus* is predominantly coastal and becomes exclusively so at the northern limits of its range. However, in many areas, most notably in West Sussex and Surrey, it penetrates many kilometres inland.

Across Ireland and Britain its distribution corresponds closely with the 5°C January mean isotherm (Doogue & Harding, 1982; Harding & Sutton, 1985). Its inland occurrence in Sussex/Surrey corresponds with an area known to have a relative mild and humid climate, which supports several outlying populations of plants with Atlantic distributions (Chater, 1984).

Habitat: *P. cingendus* is a thermophilous species requiring high summer temperatures and is intolerant of extreme winter frosts (Hopkin, 1987a). It seems to be indifferent to underlying geology and occurs on a wide variety of substrates. However, in Ireland it becomes restricted to calcareous or coastal localities towards the north of its range (Doogue & Harding, 1982).

In Ireland it occupies a wide range of synanthropic and semi-natural habitats. It is most abundant in coastal grasslands, such as damp areas of dunes, grassy sea cliffs or wherever a rough grassy sward has developed. Scrubland, hedgerows and open woodland, both coastal and inland, are also inhabited. It occurs in synanthropic sites such as gardens, churchyards and waste ground, often associated with grassy banks or overgrown walls. In Britain it is most usual to find the species in semi-natural coastal grassland, rather than dune systems, but it may be found in many similar habitats and microsites to those seen in Ireland, often on south facing slopes. One inland site in Sussex is a wet fen/carr where *P. cingendus* is associated with the hygrophilous *Ligidium hypnorum* (Chater, 1984).

Microsites: In grassland it is most easily found beneath stones (Harding & Sutton, 1985), but it can also be discovered under dead wood and mat-forming plants, within tussocks or amongst grass litter. It has been sieved from rotting seaweed on a sand/shingle beach in Cardiganshire, west Wales (Fowles, 1989). At Orford Ness, Suffolk, the most north-easterly site in Britain, the species was numerous under strandline debris, amongst reedbed and grassland litter and around disused buildings (Daws, 1994c).

Associated species: It is inevitably associated with *Philoscia muscorum*, a species for which it can be easily mistaken.

Worldwide distribution: *P. cingendus* has a strict Atlantic distribution and occurs in western coastal regions of Portugal, Spain and France (Schmalfuss, 2004). Co. Down in Ireland is the most northerly locality.

Porcellionides pruinosus (Brandt, 1833)
Metoponorthus pruinosus (Brandt, 1833)

Distinctive features: A medium sized purplish-brown woodlouse, reaching up to 12mm in length, with a characteristic blue-grey bloom. The body colour contrasts with the rather long whitish legs and the pale annulations on the antennae. The body has an obviously stepped outline allowing rapid movement.

Distribution: P. *pruinosus* is widely distributed throughout Britain and Ireland. It is common in at least central and eastern England. In many areas it undoubtedly remains under-recorded, but it does seem to become genuinely scarce in western and northern parts of Britain and Ireland. In Scotland it has been rarely recorded and appears to be mainly coastal.

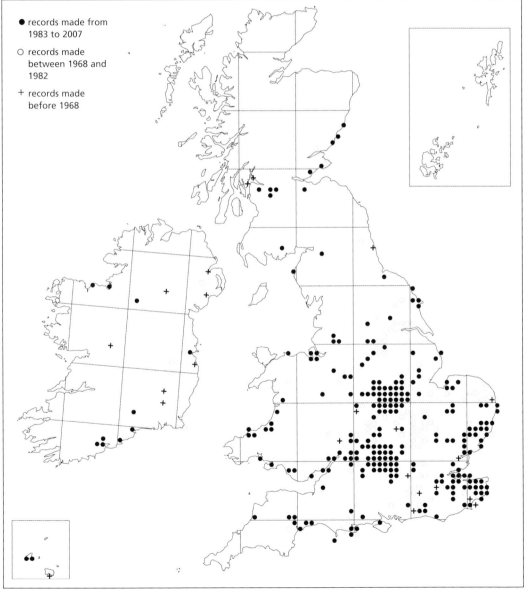

Map 52. *Porcellionides pruinosus*.

Figure 38. *Porcellionides pruinosus.* © Paul Richards.

Habitat: P. *pruinosus* is primarily associated with synanthropic habitats. Characteristically it is found within manure heaps, even those that have been transported far from farmyards. It also occurs inside and outside buildings associated with dairy farms, stables, riding schools and other animal housing. In south-eastern England it may be found in the majority of roadside manure heaps or farmyards examined, but the proportion decreases towards the north and west. In Ireland it seems to be mainly associated with stables (Cawley, 1996, 2001) and is much less frequently encountered around old cowsheds than *Porcellio dilatatus*.

This species is readily dispersed by the application of farmyard manure and is able to quickly colonise new heaps. It is able to persist in small numbers in atypical habitats, including hedgerows, road verges, woodland and fen (Daws, 1994a; Gregory & Campbell, 1995). P. *pruinosus* also inhabits churchyards, gardens and allotments, usually within compost heaps or inside glasshouses, where it is usually introduced with farmyard manure (Daws, 1994a). In Ireland populations often occur on large public refuse tips (Cawley, 1996, 2001).

Microsites: Huge populations may develop within old well-established heaps and it is usually located by digging into manure heaps, or more rarely compost heaps. Around farm buildings specimens may be found under planks, stones and pieces of manure or amongst straw and other debris. Rarely it is found amongst leaf litter and under loose bark on trees (Daws, 1994a; Gregory & Campbell, 1995). The species is gregarious and typically found clustered in tight groups.

Associated species: It is often associated with other large woodlice such as *Porcellio scaber*, *Oniscus asellus*, *Philoscia muscorum*, *Armadillidium vulgare* and the two other farmyard specialists, P. *dilatatus* and P. *laevis*.

Other notes: Harding and Sutton (1985) suggest a decline in abundance during the 20th century, due to the demise of horses for transport. Recent records indicate that it has been under-recorded and is clearly doing very well if searched for in appropriate habitats.

Worldwide distribution: Although originating in the Mediterranean region (Vandal, 1962), P. *pruinosus* has been widely introduced and occurs as a synanthrope throughout Europe and elsewhere (Schmalfuss, 2004).

Trachelipus rathkii (Brandt, 1833)
Trachelipus rathkei (Brandt, 1833)

Distinctive features: This large woodlouse, reaching up to 15mm in length, is often slatey-grey in colour, but females may be attractively marked with orange mottling. It is reminiscent of *Porcellio scaber*, but the pigment in *T. rathkii* tends to be arranged in longitudinal lines, giving the impression of pale stripes. It has been frequently confused with *Oniscus asellus*. The combination of five pairs of conspicuous pleopodal lungs and two flagella segments distinguish *T. rathkii*.

Distribution: In addition to the Northamptonshire/Huntingdonshire block noted by Harding and Sutton (1985), this species has proved to be widespread (and locally common) in Kent and South Essex. The known distribution has advanced considerably westwards to include the upper Thames

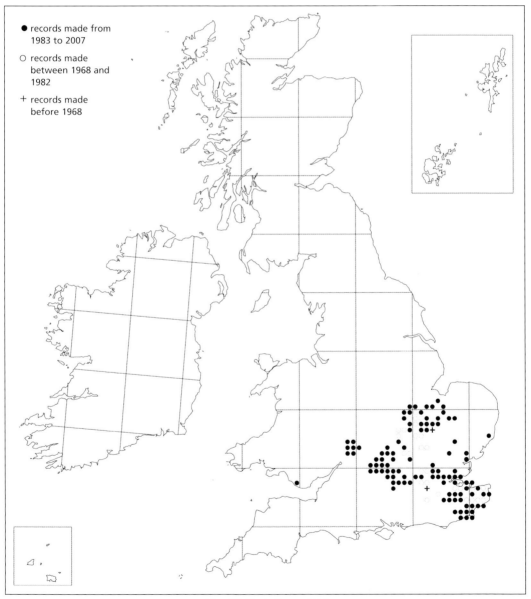

Map 53. *Trachelipus rathkii*.

Figure 39. *Trachelipus rathkii.* © Dick Jones.

Valley of Oxfordshire and Berkshire and the Severn catchment in Gloucestershire. In 2007 a site was discovered at Bridgend, south Wales near the River Ewenny (Jones, 2008). The most north-western British locality is beside the river Welland in Leicestershire (Daws, 1997). It appears to be absent from much of East Anglia, where it is probably under-recorded.

Habitat: It is apparent that the north-western extent of its range corresponds with poorly drained Jurassic Lower Lias Clays (Whitehead, 1988). Elsewhere *T. rathkii* is associated with river catchments on various geologies. The key factor seems to be the presence of soils with impeded drainage overlying heavy clays or alluvial deposits. It is tolerant of seasonal inundation (Holdich, 1988) and over much of its range is a characteristic woodlouse of riverside meadows. It also inhabits poorly drained scrub, woodland, and eroding 'soft rock' cliffs on the coast. *T. rathkii* regularly occupies synanthropic habitats, such as quarries, gravel pits, railway embankments, churchyards and gardens. Often this is in the vicinity of river flood plains, but also further afield, presumably as an introduction. In the Thames Valley, Oxfordshire it is one of the few woodlouse species pitfall trapped in ploughed arable land (Gregory, 2001).

Microsites: It can be found in a variety of typical woodlouse microsites, including under stones and dead wood, beneath bark, amongst grass litter, within tussocks (e.g. of *Deschampsia cespitosa*), amongst flood debris and in ground fissures. It is sometimes numerous, but often occurs at low densities among other species, including the superficially similar *P. scaber* and *O. asellus*.

Associated species: In addition to *P. scaber* and *O. asellus*, other associated species include *Trichoniscus pusillus* agg., *Philoscia muscorum*, *Armadillidium vulgare* and in very wet sites, *Ligidium hypnorum*. Surveys in the Netherlands suggest that *T. rathkii* competes with *P. scaber* and *O. asellus* and it tends to be more numerous in the absence of these species (Wijnhoven, 2001a).

Other notes: There is evidence that this species has recently colonised new sites in Gloucestershire (Whitehead, 1988) and the newly discovered Glamorgan locality may be a recent introduction. However, it is clear that much of the increase in range since Harding and Sutton (1985) simply reflects increased recorder activity rather than an expansion of range. *T. rathkii* remains under-recorded in many areas and further populations must await discovery in some of the larger river systems in southern and eastern England and possibly south-eastern Wales.

Worldwide distribution: This is one of the most widely distributed woodlice across central and eastern Europe (Schmalfuss, 2004), but is absent from the Iberian peninsula and the Mediterranean basin.

Additional species not mapped

Chaetophiloscia sp.

Females of a species referable to the genus *Chaetophiloscia* were collected from ornamental gardens on Tresco, Isles of Scilly (SV81) in 1985 and 1986 (Jones & Pratley, 1987). No further collections have been made since that date and in the absence of a male it has not been possible to satisfactorily name the species. This species has not been mapped. The genus *Chaetophiloscia*, in the current defined sense, is mainly found in the Mediterranean region and several species are widely distributed in southern Europe (Schmalfuss, 2004). Other allied genera are frequently reported as accidental introductions into Europe. In all probability, this is a species that has been accidentally introduced to Tresco through human activities.

Alien woodlice

This publication is primarily concerned with the native and naturalised species of woodlice that are capable of breeding outdoors in the natural climate of Britain and Ireland. Twelve additional species of woodlice that can only survive in artificial climates, such as those maintained inside glasshouses, have been recorded in Britain and Ireland. The last formally published list of alien woodlice was in Oliver and Meechan (1993), which reproduced that in Harding and Sutton (1985). This included ten species. Two additional alien species, *Styloniscus mauritiensis* and *Venezillo parvus*, are now known to occur in Britain.

In recent years collecting from glasshouses, such as those of botanical gardens, has not been popular. Much of the current information dates from the early to mid 20th Century (Edney, 1953). Many of these species were first described, new to science, from British glasshouses long before they were discovered in the wild. One species, *Miktoniscus linearis*, remains unknown in the wild. The distribution of these species has not been mapped, but 10km grid squares are given for all localities for which there have been post-1980 records.

The widespread use of insecticides in many botanical gardens may threaten the long-term survival of many alien species and may prevent further species from becoming established. However, it is worth examining well-established heated glasshouses as one never knows quite what will turn up. An exciting new prospect is the Eden Project, Cornwall, where pesticides are not routinely used. Of the species collected there recently at least two, including *Venezillo parvus* and a species of *Armadillidium* yet to be satisfactorily identified, are new to Britain.

Miktoniscus linearis (Patience, 1908)

A very small, white woodlouse reaching only up to 3mm in length. The body is strongly tuberculate and the eye composed of single black ocellus. It was described new to science from glasshouses at Kew Gardens in 1908. *M. linearis* is only known from glasshouses in England and Berlin, Germany and is so far unknown in the wild (Schmalfuss, 2004). The genus *Miktoniscus* occurs on both sides of the Atlantic and this species may originate from eastern USA (Edney, 1953).

Cordioniscus stebbingi (Patience, 1907)

This very small species, reaching 3mm, is mottled dark reddish-brown in life. The eye is composed of three ocelli and the body is covered with coarse tubercles. *C. stebbingi* was described new to science from glasshouses at Glasgow Botanic Gardens in 1907 and subsequently found in Glamorgan, the Isle of Wight and the Dublin Botanic Gardens (Edney, 1953). Recent (post-1980) records are from Glasgow Botanic Gardens (NS56), Edinburgh Botanic Gardens (NT27), Kew Gardens (TQ17), Sheffield city (SK28), where a population survived indoors in a plant pot for several years (Richards, 1995), and in 2008 Swansea University Gardens (SS69). This species originates from eastern Spain, but has been introduced and occurs as a synanthrope, mainly in glasshouses, all over the world (Schmalfuss, 2004).

Styloniscus mauritiensis (Barnard, 1936)

This species was found at Edinburgh Botanic Gardens (NT27) in August and November 1986 (Rawcliffe, 1987; Collis & Harding, 2007). The specimens were collected within peat inside plant pots. Elsewhere, *S. mauritiensis* is known from the islands of Hawaii and Mauritius (Schmalfuss, 2004).

Styloniscus spinosus (Patience, 1907)
Cordioniscus spinosus (Patience, 1907)

This small woodlouse, reaching up to 5mm in length, is similar in colour to the smaller *S. stebbingi*. It also has a tuberculate body and eyes composed of three ocelli. *S. spinosus* was described new to science from glasshouses at Glasgow in 1907 and subsequently recorded from Edinburgh (Edney, 1953). Rundle (1976) reports its presence at Kew Gardens (TQ17), but there do not appear to be any additional modern records for this species. Elsewhere it is known from Mauritius, Réunion, Madagascar and Hawaii (Schmalfuss, 2004).

Burmoniscus meeusei (Holthuis, 1947)
Chaetophiloscia meeusei Holthuis, 1947

A medium species, reaching 8mm in length. *B. meeusei* has a mottled purplish colour and a highly stepped pereon/pleon outline. It was described new to science from glasshouses at Kew Gardens in 1947 (Edney, 1953). Although re-found at Kew (TQ17) by Rundle (1976), there have not been any subsequent records. Elsewhere it appears to have a very wide distribution in the tropics, including Hawaii, Brazil and Taiwan (Schmalfuss, 2004).

"Setaphora" patiencei (Bagnall, 1908)
Chaetophiloscia patiencei (Bagnall, 1908)

In life this very small woodlouse, reaching up to 3mm in length, is violaceous brown mottled with white. It was described new to science from glasshouses at Kew Gardens by Bagnall in 1908 and subsequently found at Winlanton, Durham (Edney, 1953). There are no modern records for this species. The taxonomy of this species is poorly understood, but it is clear that it does not belong to the genus *Setaphora* within its present definition (Schmalfuss, 2004), hence the quotation marks. Elsewhere this species is known from Mauritius and Réunion.

Trichorhina tomentosa (Budde-Lund, 1893)

This is a small white species, reaching 5mm in length. It has an distinctive oval body outline and has eyes composed of a single black ocellus. In recent years (post-1980) *T. tomentosa* has been recorded from Glasgow Botanic Gardens (NS56), Tolcross Gardens, Glasgow (NS66), Belfast Botanic Gardens (J37), the Natural History Museum, Oxford (SP50), where it occurs in a heated Cockroach cage, the Eden Project, Cornwall (SX05) and most recently from a heated reptile house in Somerset (ST04). This species reproduces parthenogenetically and males are unknown. In the wild it is known from tropical Central and South America, but has been introduced to glasshouses worldwide (Schmalfuss, 2004).

Reductoniscus costulatus Kesselyák, 1930

This very small woodlouse, reaching 2mm, is capable of rolling into a sphere when disturbed. It was first recorded from Kew Gardens (TQ17) in 1947 and has been rediscovered there in recent years. *R. costulatus* is well known inside other glasshouses across Europe and occurs outdoors in Seychelles, Mauritius, Malaysia and the Hawaiian Islands (Schmalfuss, 2004).

Figure 40. *Reductoniscus costulatus*. © Paul Richards.

Venezillo parvus (Budde-Lund, 1885)

This is a larger ball-rolling species, reaching 5mm in length. The background colour is dark brown, but a pale dorsal band is apparent and two longitudinal rows of pale patches are present on each pereonite. Large numbers of this species were collected from the Eden Project, Cornwall, (SX05) in 2005, but only recently identified (Gregory, 2009). This species is widespread in tropical and sub-tropical regions (Schmalfuss, 2004), but also introduced into glasshouses in Europe (personal communication, S. Taiti).

Agabiformius lentus (Budde-Lund, 1885)

This is a small species reaching 6mm. The strongly arched body is pale with three broad ill-defined stripes. *A. lentus* was reported by Sutton (1972) to be our most widespread glasshouse alien, but there appear to be no modern (post-1980) records. This species originates from the Mediterranean coasts, but has been introduced to many other parts of the world by human activities (Schmalfuss, 2004).

Nagurus cristatus (Dollfus, 1889)

A medium sized species, to 10mm. The body has a yellowish background with four irregular brownish longitudinal stripes. This is a parthenogenetic species. In Britain *N. cristatus* has been recorded from Northumberland in 1965 (Sutton, 1972). There have been no additional records. It has been widely introduced throughout the tropics (where it occurs outdoors) and it occurs as a synanthrope inside glasshouses in temperate regions (Schmalfuss, 2004).

Nagurus nanus (Budde-Lund, 1908)

A small species, to 4.5mm. A single specimen was collected from a heated glasshouse in Belfast Botanic Gardens in 1911 (Sutton, 1972). There have been no subsequent records. *N. nanus* has been widely introduced throughout the tropics where it occupies disturbed habitats (Schmalfuss, 2004).

Habitat accounts

The following section is based on the Habitat Accounts in *Woodlice in Britain and Ireland* (Harding and Sutton, 1985). This describes woodlice assemblages recorded from major habitat types listed within the Habitat Classification printed on the reverse of the non-marine isopod recording card in use from 1970 until 1982. The habitat accounts have been updated, where appropriate, in light of subsequent observations.

Five species (*Oniscus asellus*, *Porcellio scaber*, *Trichoniscus pusillus* agg., *Philoscia muscorum* and *Armadillidium vulgare*) occur in all major habitat types. The first three typically occur in large numbers across much of the Britain and Ireland, *P. muscorum* becomes predominantly coastal in northern Scotland, while *A. vulgare* is only common in the south and east. Records of these five eurytopic species combined make up 77% of all records received by the survey. Other species are much less frequently encountered. Woodlice occur in a wide range of habitats, but for convenience these can be considered as three broad categories: coastal habitats, inland semi-natural habitats and man-made (synanthropic) habitats.

Coastal habitats

Coastal habitats generally support the most diverse woodlice assemblages, including a specialist fauna associated with the supralittoral zone. Strandline debris, such as seaweed and driftwood, and soft cliffs subject to frequent landslips provide plenty of shelter for a great variety of woodlice species. For this reason coastal habitats are considered in more detail than other habitat types.

Salt marshes

Salt marsh forms where silt accumulates in shallow tidal water, in the estuaries of rivers or behind dune systems and shingle ridges. Most species of woodlice appear to be unable to survive frequent inundation in saline water, so that salt marshes are generally not good areas for them. Despite this Harding and Sutton (1985) record thirteen species. Woodlice occur on salt marshes almost exclusively in the strandline, although occasional specimens (mainly *Ligia oceanica*, *Porcellio scaber* and *Armadillidium album*) may be found under isolated pieces of drift material in the upper parts, mainly in the areas dominated by *Festuca rubra* and *Juncus* sp.

The species most frequently recorded on salt marshes (in decreasing frequency) are *P. scaber*, *L. oceanica* (usually associated with wood piles and solid structures), *Oniscus asellus*, *Philoscia muscorum*, *Armadillidium vulgare* and *Trichoniscus pusillus* agg. Also recorded are *Porcellionides cingendus*, *Cylisticus convexus* and *Oritoniscus flavus*. One rare species *Miktoniscus patiencei* is characteristic of salt marsh strandlines, but two other scarce species, *A. album* and *Trichoniscoides saeroeensis* may occasionally be found. *M. patiencei* and *Platyarthrus hoffmannseggii* have been recorded regularly in association with ants just inland from the upper strandlines.

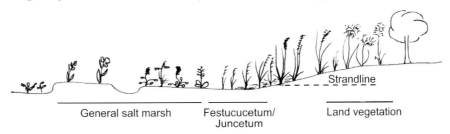

Figure 41. Diagrammatic profile of salt marsh.

Sand dunes

Harding and Sutton (1985) record twenty species from sand dunes, but only *Armadillidium vulgare*, *Philoscia muscorum* and *Porcellio scaber* are widespread and common. *P. scaber* is typically very numerous on dune systems and may be the only species present in the yellow dune zone. At some sites, and in damp or windless conditions, *A. vulgare* and *P. muscorum* may also be present. *P. muscorum* is characteristic of tall, stabilised vegetation, particularly in dune slacks. At some northern and western dune sites *P. muscorum* can be the most abundant species. *A. vulgare* is usually associated with areas dominated by marram *Ammophila arenaria* or with shorter grassland with scattered marram. In Ireland *Porcellionides cingendus* occurs on dunes, mainly in areas of dense grassy swards and in damp locations.

Armadillidium album is found among sea drift material in the strandline, especially where this is buried at depths of 20-30cm in sand. *A. vulgare* and *P. scaber* are also occasionally recorded in the strandline. *Platyarthrus hoffmannseggii* is recorded in association with ant nests in stabilised zones. *Trichoniscus pusillus* agg. is recorded throughout, usually in association with damp hollows, mossy areas, large pieces of concrete or decaying wood, but mainly in dune slacks. *Oniscus asellus* is recorded almost invariably in association with ruined buildings, decaying wood and other debris and rubbish.

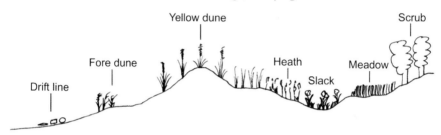

Figure 42. Diagrammatic profile of sand dune.

Shingle beaches

Shingle shores are widespread in the British Isles, forming substantial lengths of coastline along the south and east coasts of England. Shingle is often mixed with sand in varying quantities and, on the shores of some estuaries, with silt. Shingle banks and ridges, when fully developed, form wide, fairly flat-topped features with a zonation of vegetation. On the seaward side of zone A the beach is devoid of vegetation. Above this sparse strandline vegetation begins to colonise (zone A) which gives way to established maritime grassland with herbs (zone B). In some places shingle builds up as an almost unvegetated bank against coastal grassland or other terrestrial vegetation.

The foreshore of shingle beaches, between mean high water mark (HWMN) and the extreme high water mark of spring tides (EHWMOS), where vegetation is absent, may seem an improbable habitat for woodlice, but this is the typical situation in which *Stenophiloscia glarearum* is found (Harding, Cotton & Rundle, 1980), and from which *Halophiloscia couchii* has also been recorded.

Several species are recorded in zone A among the strandline and pioneer vegetation of sea rocket *Cakile maritima*, sea sandwort *Honkenya peploides* or sand couch-grass *Elytrigia juncea*. These include *Porcellio scaber*, *Armadillidium vulgare*, *Cylisticus convexus*, *Oniscus asellus*, *Philoscia muscorum* and occasionally *Porcellionides cingendus* or *H. couchii*. Deeper in the shingle, small species have been found among damp soil derived from drift material, including *Trichoniscus pusillus* agg., *T. pygmaeus*, *Trichoniscoides saeroeensis* and *Miktoniscus patiencei*. Similar assemblages are recorded in zone B, but a more humus-rich soil develops deep in the shingle, in which *Haplophthalmus mengii* seg., *Androniscus dentiger* and *Buddelundiella cataractae* occur.

Areas of shingle beaches and vegetated ridges have not been adequately sampled because special techniques are required, such as pitfall traps for *S. glarearum* and sieving and floatation for the deep-dwelling species, especially *B. cataractae*.

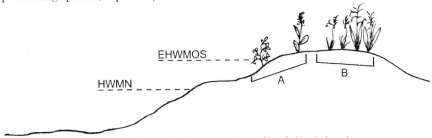

Figure 43. Diagrammatic profile of shingle beach.

Boulder beaches

On exposed coasts where physical and geological factors permit, storm beaches have formed which are composed primarily of rounded stones and pebbles. In some instances, these have built up against cliffs or have formed round-topped banks reaching to or above high water mark. Nothing is known of the fauna below extreme high water mark because of the difficulty of sampling among the stones.

Just above extreme high water mark, the stones are stabilised by a sparse vegetation of strandline plants (zone A). In the top layers (h1) little is found, except for occasional specimens of *Porcellio scaber* and *Armadillidium vulgare*. Deeper down, where the humidity is higher and where there are fragments of soil (h2), *Oniscus asellus*, *Trichoniscus pusillus* agg., *Androniscus dentiger* and occasionally *Halophiloscia couchii* occur, mainly in crevices. Deeper again (h3), where more soil has accumulated, soil-dwelling species occur, including *Haplophthalmus mengii* seg., *Trichoniscus pygmaeus* and *Buddelundiella cataractae*.

The fauna of zone B, further above, is usually similar to that of vegetated cliffs or any other coastal grassland, depending on the conformation of the coastline.

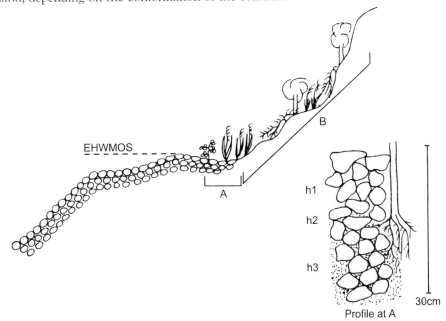

Figure 44. Diagrammatic profile of boulder beach.

Erosion banks

Where low-lying land gives way to a rocky shore the soil and vegetation are often eroded to leave a low, steep slope to the foreshore, often only 1-3m high. The erosion banks thus formed often have a distinct zonation of vegetation. The uppermost zone, C, can be grassland, woodland, scrub or any other typical terrestrial vegetation. Zone B is strandline vegetation comprising such species as oraches *Atriplex* spp., docks *Rumex* spp. and sow-thistles *Sonchus* spp. The upper littoral zone A is usually devoid of vegetation.

Ligia oceanica is almost exclusively found in zone A, but occasionally *Halophiloscia couchii* also occurs here. In zone B *Porcellio scaber*, *Armadillidium vulgare* and *Cylisticus convexus* are most frequently found under loose rocks or larger pieces of strandline debris. Smaller species occur mainly under rocks lying on eroded soil and among the roots of strandline plants (h1). Close to the surface *Trichoniscus pusillus* agg. and *Androniscus dentiger* are most frequent, but deeper down, under the larger stones and deep among roots, *Trichoniscus pygmaeus* and *Trichoniscoides saeroeensis* occur. The same larger species occur in zone C but, where large stones are embedded in the soil (h2), a greater variety of small soil-dwelling species is recorded, with *Haplophthalmus mengii* seg. often the most frequent. This severely limited and fragile habitat is where *Metatrichoniscoides celticus* occurs in south Wales.

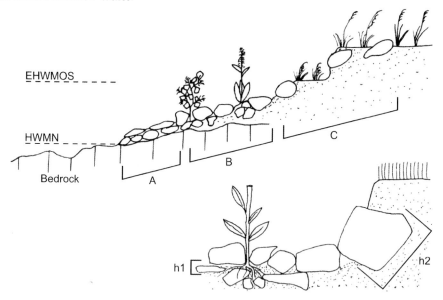

Figure 45. Diagrammatic profile of erosion bank.

Vegetated cliffs

Vegetated cliffs are typically formed from a softer substrate than that of near-vertical hard rock cliffs. They are characteristically sloped or terraced as a result of frequent slumps and landslips caused by erosion, either at the cliff base by wave action, from above by rain or by groundwater seepage through the cliff.

This basic habitat covers a wide variety of different types depending on the underlying geology, slope, aspect and exposure, and also the geographical area. The splash zone usually has a vegetation of lichens, scurvy-grass *Cochlearia* spp. and sea campion *Silene uniflora*, with further up the cliff, species such as rock samphire *Crithmum maritimum*, greater sea-spurrey *Spergularia media*, thrift *Armeria maritima*, rock sea lavender *Limonium binervosum* agg. and sea plantain *Plantago maritima*. Transition into a dense grassy sward is often gradual and may, in fact, be into low scrub of scrambling shrubs (e.g. brambles *Rubus* spp., roses *Rosa* spp.) rather than grass. Where cliffs are grazed the grassland is often short so that woodlice are unable to shelter in the grass litter and tend to occur in crevices, under stones and in the soil.

The fauna of the splash zone is very rich with many soil-dwelling species occurring in the damp loamy soil derived from drift material and the litter of mat-forming species such as sea campion. *Androniscus dentiger*, *Haplophthalmus mengii* seg., *Trichoniscoides saeroeensis*, *Trichoniscus pusillus* agg. and *Trichoniscus pygmaeus* have all been recorded. The foliage of mat-forming plants provides day shelter sites for *Porcellionides cingendus* and *Eluma caelatum*, as well as for the much commoner *Porcellio scaber* and *Armadillidium vulgare*. *Ligia oceanica*, and occasionally *Halophiloscia couchii*, also occur in the splash zone.

Grassy swards further up the cliffs support *A. vulgare*, *Philoscia muscorum*, *P. scaber* and, more rarely, *Armadillidium depressum*, *A. nasatum* and *A. pulchellum*. The last species is found in Wales and Scotland mainly under flat stones, often in association with ants. In such situations in the south, *Platyarthrus hoffmannseggii* is common. *Acaeroplastes melanurus* inhabits sparsely vegetated cliffs and during inclement weather seeks refuge (with *P. scaber*) amongst lichens covering exposed rocks. The fauna of scrub on cliffs is similar to that of shaded situations inland, but species dependent on base-rich soils occur in otherwise unexpected sites on acidic rocks, probably as a result of wind-borne sea spray maintaining more basic soil conditions.

Androniscus dentiger, *Cylisticus convexus* and *Trichoniscoides albidus* seem to be often associated with seepages on cliffs, particularly those cliffs composed of clay or other soft rocks. The first two species are most frequently encountered on the cliff face during darkness. *T. albidus* often occurs between layers of clay or beneath large stones and dead wood where these are present. *Miktoniscus patiencei* is found in deep, friable soil under dense fescue (*Festuca* spp.) grassland, usually in sheltered clefts in the cliff face, only a metre or two above the rocky shore edge where only lichens grow.

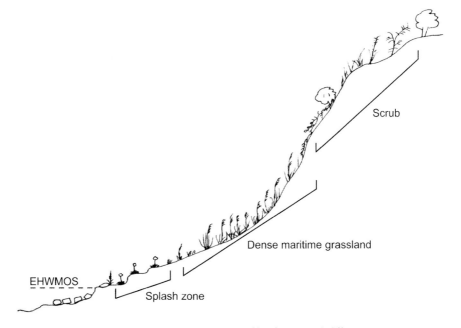

Figure 46. Diagrammatic profile of vegetated cliff.

Steep soil-less cliffs

Steep cliffs, which lack soil and vegetation, are inhabited by few species of woodlice. Throughout the British Isles *Ligia oceanica* is common on such cliffs wherever sampled. On exposed southern and western coasts *Halophiloscia couchii* is also found, but its nocturnal habits and the inaccessibility of its day shelter sites can make this species difficult to find.

More species are encountered where a boulder talus has formed at the base of the cliff, especially if drift material had decayed to form soil among the talus. The fauna is the same as that described for erosion banks.

On very exposed cliffs the effects of maritime influences are seen often as much as 100m above high water mark. For example, *L. oceanica* can be found at the top of several high cliffs and *Trichoniscoides saeroeensis* can be found among the roots of vegetation or beneath deeply embedded stones on cliff tops.

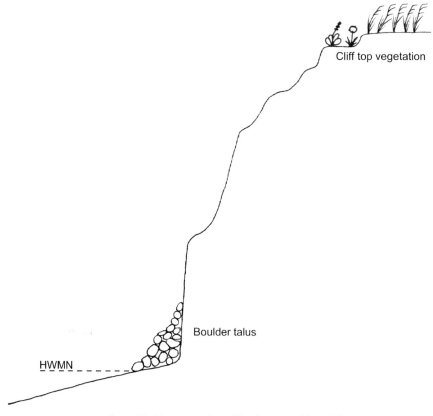

Figure 47. Diagrammatic profile of steep soil-less cliff.

Inland semi-natural habitats

Subterranean habitats

Few woodlice records have been made from subterranean habitats, such as caves and mines, and species occurrence is probably seriously under-recorded. The British Isles does not hold a specialist subterranean (troglobitic) fauna. Harding and Sutton (1985) report eleven species occurring in subterranean situations, but these were generally collected from the cave threshold. Here *Oniscus asellus* is the most frequently reported species, but also other eurytopic species including *Armadillidium vulgare*, *Philoscia muscorum*, *Porcellio scaber*, *Trichoniscus pusillus* agg. and *Haplophthalmus danicus*. On the sea-shore *Halophiloscia couchii* and *Ligia oceanica* may be found at the threshold of caves in sea cliffs.

Two species, *Androniscus dentiger* and *Trichoniscoides saeroeensis*, are able to maintain viable populations deep inside cave and mine systems, but since they also occur above ground they cannot be considered to be true troglobitic species. Specimens have been recorded deep underground from under stones, in shallow water films or within waterlogged timbers. Occasionally *Trichoniscus pygmaeus* has been found associated with these species, but often close to the threshold. *Cylisticus convexus* has been recorded from tunnels, and this species, together with *Androniscus dentiger*, has been recorded from underground drains, sewers and buried cable trunking in urban areas. These species are likely to be very under-recorded in these, often unsavoury, situations.

Wetlands

Wetlands are transitional habitats between 'terrestrial' dry land and 'aquatic' open water. Four types of wetland are described by Harding and Sutton (1985), including fen, bog and carr (the fourth type, salt marsh, is described under Coastal Habitats). These terms are poorly defined. Fens are often taken as those wetlands that are groundwater fed (but often loosely applied to those fed by calcareous groundwater). Bogs are predominantly fed by rainwater (but base-poor fens are often referred to as 'bogs'). Carr is waterlogged woodland, typically developed on fen peat.

Harding and Sutton (1985) recorded twenty species from wetland habitats, but only four species, *Ligidium hypnorum*, *Haplophthalmus mengii* seg., *Trichoniscoides albidus* and *Trachelipus rathkii*, can be considered to be characteristic of wetland habitats. The soil-dwelling trichoniscids are often found on the underside of large stones or pieces of dead wood partly embedded into soil. *Trichoniscus pusillus* agg. is common in all wetland types. *Haplophthalmus mengii* seg. is more common in fens and bogs than in carr, while *Trichoniscus pygmaeus* is recorded more commonly in bogs than in fens or carr. *Trichoniscoides albidus* is more frequent in fens, especially those associated with river floodplains, but also occurs in carr.

Tussocky vegetation is the most frequently recorded marsh location for *Philoscia muscorum*, but this species along with *L. hypnorum*, *Porcellionides cingendus* and *Trachelipus rathkii* are characteristic of fen and carr litter. *Haplophthalmus danicus* can also be numerous among wet litter and moss in fens. Dead wood and artificial piles of dead vegetation, such as that cleared from fen dykes, provided sites for *Oniscus asellus*, *Porcellio scaber* and *H. danicus*. A final suite of species, including *Armadillidium vulgare*, inhabits drier sites such as dykeside banks.

Grasslands

Grasslands are one of the most frequently recorded major habitat types. Twenty-eight species were recorded by Harding and Sutton (1985). The most frequently recorded are *Oniscus asellus*, *Porcellio scaber*, *Philoscia muscorum*, *Trichoniscus pusillus* agg. and *Armadillidium vulgare*. *Platyarthrus hoffmannseggii*, *Androniscus dentiger*, *Trichoniscus pygmaeus*, *Haplophthalmus mengii* seg. and *Porcellionides cingendus* also occur regularly. Woodlice tend to be more abundant in ungrazed and lightly grazed grasslands, but a few species, such as *Armadillidium pulchellum*, are tolerant of heavily grazed swards. Grassland is an important habitat for *P. muscorum*, *P. hoffmannseggii* and *Trachelipus rathkii* with above average occurrences of each reported by Harding and Sutton (1985).

It can be difficult to find species in grassy habitats. Shelter sites, such as stones or dead wood, can be useful but if these are not present it is necessary to search through the sward, tussocks or grass litter. It is apparent that different grassland types support distinct assemblages of woodlice species.

Lowland calcareous grassland is important for *P. hoffmannseggii*, which is normally found in association with ants, particularly under stones or within anthills. *Trichoniscoides helveticus* is characteristic of scrubby lowland calcareous grassland in England and may be sorted from soil or found beneath large stones partially embedded in soil. *P. muscorum* is widespread and abundant in lowland calcareous grassland, as to a lesser extent is *A. vulgare*. Upland calcareous grassland is a key habitat for *A. pulchellum*, which, in England, is mainly associated with screes and short turf swards on the northern Carboniferous limestones.

In the English lowlands *T. rathkii* is found mainly in ill-drained grassland on clayey soils, especially those associated with river floodplains. This is also a characteristic habitat for *H. mengii* seg. and *Trichoniscoides albidus*, which may be found beneath stones and dead wood or amongst rubble in damp hollows and at ditch margins. The fauna of damp seasonally inundated grassland is more akin to that of wetlands than to other grassland types.

Scrublands

Scrub is generally a transitional habitat between grassland and woodland and it shares, therefore, many similarities with each. Open scrub with grass and herbs generally supports the greatest woodlice interest. Of the twenty-four species recorded by Harding and Sutton (1985), four are frequent: *Oniscus asellus*, *Trichoniscus pusillus* agg., *Philoscia muscorum* and *Porcellio scaber*. Other species that occur regularly are *Armadillidium vulgare*, *Trichoniscus pygmaeus*, *Androniscus dentiger*, *Porcellionides cingendus* and *Haplophthalmus mengii* seg.. Harding and Sutton (1985) reports that three species are recorded in scrubland with above average frequency: *T. pusillus* agg., *Armadillidium pictum* and *Oritoniscus flavus*.

Woodlands

Woodlands are another frequently recorded major habitat. Although twenty-six species are recorded by Harding and Sutton (1985), only four are frequent: *Oniscus asellus*, *Trichoniscus pusillus* agg., *Porcellio scaber* and *Philoscia muscorum*. Other species that occur regularly are *Armadillidium vulgare*, *Trichoniscus pygmaeus*, *Androniscus dentiger*, *Haplophthalmus danicus*, *Haplophthalmus mengii* seg. and *Ligidium hypnorum*. Generally woodlice avoid dense shade and favour open woodland where herbs and grass are present.

Woodland holds a distinctive fauna and *Armadillidium pictum*, *H. danicus*, *L. hypnorum* and *T. pusillus* agg. all had above average numbers of records by Harding and Sutton (1985). Leaf litter and the damp soil/litter interface are important habitats for woodlice in woodland. The frequent occurrence of *O. asellus* and *T. pusillus* agg. reflects the importance of leaf litter as a moisture-retaining habitat since both species are associated with more damp situations than, for example, *P. scaber*. Large pieces of fallen rotting timber, particularly when partly embedded into soil or lying in the bottom of damp hollows or ditches, can also be productive for many species such as *L. hypnorum* and small trichoniscids, including *H. danicus* and *Trichoniscoides albidus*.

Woodland is an important habitat for three rare species: *A. pictum*, *H. montivagus* and *T. helveticus*. Since the publication of Harding and Sutton (1985) it has become apparent that *A. pictum* may occur in deadwood niches on tree boles well above ground level. *T. helveticus* and *H. montivagus* typically occur under large partially embedded stones, the latter often in damp situations.

Acid heathlands/moors

This is another poorly recorded habitat. Harding and Sutton (1985) report the occurrence of fifteen species, of which only the four eurytopic species, *Oniscus asellus*, *Porcellio scaber*, *Trichoniscus pusillus* agg. and *Philoscia muscorum* are at all frequent. Other species included *Armadillidium vulgare*, *Armadillidium pulchellum*, *Porcellio spinicornis* and *Porcellionides cingendus*. Many of the other species, and particularly *A. vulgare* and *P. spinicornis*, are recorded in association with walls, buildings and other synanthropic features.

This habitat is important for only one species, *A. pulchellum*, which occurred with above average frequency, mainly among screes from limestone outcrops and in small calcareous pockets in acidic grassland.

Man-made (Synanthropic) habitats

Many species of woodlice are able to readily colonise sites disturbed by human activities. Some synanthropic species occur naturally in habitats on the coast (e.g. *Cylisticus convexus* and *Androniscus dentiger*) or on river flood plains (e.g. *Trachelipus rathkii*) that are subject to natural disturbance, and they readily take to humanly disturbed (synanthropic) habitats. Other species, such as *Eluma caelatum* and *Buddelundiella cataractae*, are probably ancient introductions and are strongly associated with human activity. Synanthropic habitats have tended to be neglected, in favour of semi-natural sites. However, if a balanced picture of our woodlice fauna is to be achieved it is important that synanthropic habitats are also surveyed.

Buildings

Buildings and their immediate surroundings have been well surveyed. All the eurytopic species occurred in association with buildings. Not surprisingly, the faunas associated with buildings and gardens have many similarities.

Several species are frequently recorded inside occupied dwelling houses. *Porcellio scaber* is the woodlouse most frequently recorded indoors, but *P. spinicornis*, especially in western areas, and *Armadillidium depressum* are also characteristic species. The latter may be abundant inside older dwellings at night. *Armadillidium vulgare* and *A. nasatum* also occur indoors to a lesser extent. Sometimes *Oniscus asellus* may be found associated with rotten timber window frames.

Ruins and outbuildings provide records of *P. spinicornis*, *P. dilatatus*, *Porcellionides pruinosus* and *A. nasatum*. Farm outbuildings, barns, stables, etc, are important sites for *P. pruinosus*, *P. dilatatus* and *Porcellio laevis*. Specimens are normally found among old straw bedding and stable litter, especially in neglected corners. Due to their often unpleasant nature these habitats have been rather neglected by most surveyors. The few targeted surveys have demonstrated that *P. dilatatus* is a characteristic and predictable inhabitant of dairy farms and stables (Daws, 1994a; Cawley, 1996, 2001) and all three species are probably under-recorded.

The outside walls of buildings have provided many records, but there is some overlap with records from gardens. In damper situations, particularly at the wall/soil interface, *Trichoniscus pusillus* agg., *Philoscia muscorum*, *Androniscus dentiger* and *Trichoniscus pygmaeus* have been recorded. Higher up on walls *Oniscus asellus*, *P. scaber* and *P. spinicornis* are recorded frequently. Although hiding in crevices by day, the few night-time records demonstrate that these species are often abundant on the wall surface by night.

Glasshouses have a particularly distinctive fauna with many alien species occurring only in heated glasshouses, particularly those in botanic gardens. *A. nasatum* was formerly widely recorded in commercial glasshouses. Although recent records have been relatively scarce, surveys of gardens centres and botanic gardens indicates that it remains a characteristic species of both unheated and heated glasshouses.

Gardens

Domestic gardens have been generally well recorded. Twenty-five species are recorded by Harding & Sutton (1985), including almost every native and naturalised species, with the exception of *Armadillidium pictum*, *Armadillidium pulchellum* and the exclusively coastal species. Churchyards hold a very similar fauna to gardens. The five eurytopic species predominate: *Oniscus asellus*, *Porcellio scaber*, *Philoscia muscorum*, *Trichoniscus pusillus* agg. and *Armadillidium vulgare*.

Several features of gardens provide important habitat for woodlice. Rockeries are an important habitat for *Androniscus dentiger*, *Armadillidium depressum*, *Armadillidium nasatum*, *Porcellio laevis* and *Trichoniscoides albidus*. Several species have been recorded from flowerbeds: *A. dentiger*, *Platyarthrus hoffmannseggii*, *Porcellionides pruinosus*, *Cylisticus convexus*, *Porcellio dilatatus* and *Trachelipus rathkii*. Compost, refuse and manure heaps, including those on allotments, are important for *P. pruinosus*, *P. dilatatus* and *P. laevis*, and also *A. dentiger*, *Eluma caelatum* and *Haplophthalmus danicus*. Walls in gardens frequently provide sites for *Porcellio spinicornis* and *A. depressum*. In western areas *Porcellionides cingendus* is occasionally recorded from gardens.

The richness of the soil fauna of gardens is demonstrated by the diversity of soil-dwelling trichoniscid species recorded. Carefully sorting friable soils at the base of walls may reveal soil-dwelling species such as *Haplophthalmus mengii* and *Trichoniscoides sarsi*. *Buddelundiella cataractae* has been collected from among peaty soil beneath paving stones, associated with *Trichoniscus pygmaeus* and *H. mengii* seg.

Churchyards

Churchyards, if they are not too neat and tidy, can provide a wide array of habitats able to support a broad suite of woodlice species. All species found in gardens and buildings may be found in churchyards. The large proportion of records of many species made from churchyards is an example of recorder bias. Since they are open to the public, churchyards are easy to survey and can provide a good indication of the species inhabiting the local (otherwise private) built-up area.

Well-rotted compost heaps can support *Haplophthalmus danicus*, *H. mengii*, *Porcellionides pruinosus*, and more rarely *Porcellio laevis* or *P. dilatatus*. Large stones partly embedded into soil, including piles of broken gravestones, heaps of rubble and friable soil at the base of walls can be extremely productive for soil-dwelling trichoniscids. Frequent species include *Androniscus dentiger*, *H. mengii* and *Trichoniscus pygmaeus*, but it is in similar locations that the rare *Trichoniscoides sarsi* has been found. *Platyarthrus hoffmannseggii* may be found in ants' nests under stones or beneath flower vases. Old churchyard walls, especially if composed of limestone, and piles of stones or slates may hold *Porcellio spinicornis* and *Armadillidium depressum*. Lifting capping stones or night-time torch surveys are the easiest ways to survey for these last two species.

Brownfield sites

Brownfield sites, such as waste ground, disused quarries and pits, old railway lines, disused airfields and post-industrial sites, have been substantially altered by human activity. They share a number of features with coastal habitats, such as low fertility, sparse vegetation and regular disturbance. As with semi-natural coastal sites, they can support a wide array of woodlice species.

Harding and Sutton (1985) record twenty-nine species from 'waste ground'. Twice the number of records were made from sites with 'more than 25% vegetation cover' than from sites with 'less than 25% of vegetation cover'. The eurytopic species, *Oniscus asellus*, *Porcellio scaber*, *Philoscia muscorum*, *Trichoniscus pusillus* agg. and *Armadillidium vulgare*, are well represented. Other species which occurred regularly are *Androniscus dentiger*, *Platyarthrus hoffmannseggii*, *Trichoniscus pygmaeus* and *Cylisticus convexus*.

A few species are recorded with above average frequency, constituting possibly a characteristic waste ground fauna: *A. dentiger*, *C. convexus*, *Armadillidium depressum*, *Eluma caelatum*, *Trachelipus rathkii* and *Trichoniscoides albidus*. Recent observations suggest that *Buddelundiella cataractae* should be added to this list. Many of the niches are similar to those described for 'soft' vegetated cliffs and for supralittoral habitats, such as sparse strandline vegetation. For example, small soil-dwelling trichoniscids, such as *A. dentiger*, *B. cataractae*, *Haplophthalmus mengii* and *Trichoniscus pygmaeus* tend to be found beneath large stones or several centimetres below ground, where damp peaty soil has accumulated within stony ground or ballast.

Arable

Arable land, which includes cereal, root and fodder crops, and strictly speaking also temporary grass leys, are poorly surveyed habitats. Cereal fields are the most adequately surveyed type. Harding and Sutton (1985) report eighteen species, but highlight the problem that many records attributed to arable fields are from marginal habitats adjacent to arable land, such as hedges, roadside verges, embankments or dry stone walls, or from manure heaps, and therefore reflect the fauna of these marginal habitats rather than of arable crops themselves.

The species associated with these marginal habitats included *Porcellio spinicornis*, *Haplophthalmus danicus*, *Porcellionides pruinosus*, *Armadillidium nasatum* and *Platyarthrus hoffmannseggii*. These species are associated particularly with walls, and stones and rubbish on roadside verges. *P. pruinosus*, *Porcellio dilatatus* and *P. laevis* have been recorded in dung heaps or associated with dung spread on fields.

The common eurytopic species, *Oniscus asellus*, *Porcellio scaber*, *Philoscia muscorum*, *Trichoniscus pusillus* agg. and *Armadillidium vulgare* occurred in proportions similar to those recorded for waste ground. These and *Trachelipus rathkii*, one of the few species to be taken in pitfall traps set in arable fields, are able to penetrate deep into arable fields. However, limited sampling of arable crop fauna suggests that few woodlice inhabit the vegetation or soil of arable fields permanently. *Oritoniscus flavus* has been recorded in grass ley in Ireland.

Biogeography

Origins of the British and Irish fauna

Isopods are poorly represented in the fossil record. Unlike plants (which have pollen) and some invertebrate taxa (such as snails, which have hard shells) they do not preserve well. Thus, comments on the origin of our current fauna are somewhat speculative. The last glacial (Devensian) period peaked around 20,000 years ago and most of Britain and Ireland would have been covered with thick ice-sheets, several hundred metres in depth. Even though southern areas were free of ice the inhospitable permafrost conditions would have eradicated the great majority of the species we know today. It is debatable whether exceptions, such as the stygbiotic asellid *Proasellus cavaticus*, may have been able to survive deep underground within cave systems (Proudlove, Wood, Harding, Horne, Gledhill & Knight, 2003). All other species were forced southwards; the most frost-sensitive ones into isolated refugia in southern Europe.

The majority of our isopod fauna, if not all, must therefore have recolonised after the ice-sheets began to retreat. Sub-arctic conditions may have prevailed until about 10,000 BP, but some of our most frost-hardy woodlice species, such as *Trichoniscus pusillus* agg., could have colonised southern England at an early stage. Isostatic changes in sea level immediately after the post-glacial period may have been an important factor in allowing the colonisation of Ireland and south-western Britain. There is evidence of a low-lying land bridge connecting south-west England with Ireland, but this was completely severed by 13,000 BP (Lambeck, 1996). This is far too early for frost-sensitive species such as *Porcellionides cingendus* to have colonised Ireland via England. Additionally, this land bridge would have been inundated regularly with seawater, making it a hostile environment for any plants or animals attempting a crossing (Whitehouse, 2006).

It is thought that the connection between Britain and continental Europe was finally inundated by rising sea levels around 8,500 BP. This leaves a narrow window of opportunity, some 1,500 years, following the onset of early Holocene warming (c. 10,000 BP) for less hardy species to 'walk' across the land bridge from mainland Europe. There are likely to have been a number of potential routes for colonisation. One possible mechanism was the massive proto-Rhine which bisected the land bridge and carried the combined flow of the Thames, Rhine, Meuse, Scheldte, Somme and Seine, before discharging into the Atlantic Ocean (Gupta, Collier, Palmer-Felgate & Potter, 2007). Rather than being a barrier to dispersal, it is more likely that this massive river system transported species out of Europe (e.g. via flood debris) and allowed penetration into southern England (e.g. via the Thames Valley).

The relatively early loss of the land bridge connection to mainland Europe means that in Ireland and, to a lesser extent, Britain a number of species must have arrived after its inundation. One natural dispersal mechanism to reach our shores is to raft across the English Channel and/or the Irish Sea by clinging to vegetation mats, driftwood trees and other debris washed out to sea from European rivers. The isopod fauna of the British Isles includes a large proportion of 'expansive' species that are able to disperse along lowland river valleys and coastlines in this way (Vandel, 1960). Examples include some of our more local species; *Haplophthalmus danicus*, *Trichoniscoides albidus*, *Ligidium hypnorum* and *Trachelipus rathkii*.

Other woodlice, such as *Eluma caelatum*, exhibit erratic and/or discontinuous distribution patterns that are not easily explained by natural dispersal mechanisms, but begin to make sense if unintentional introduction by Mesolithic traders is considered (Corbet, 1962). Barber and Jones (1996) consider our millipede fauna to be a product of chance introduction over many millennia. Recently, Welter-Schultes (2008) demonstrated that terrestrial snails were accidentally transported via Bronze Age ships around the eastern Mediterranean. These dispersal mechanisms are equally plausible for our isopod fauna.

The importance of human activity

There are two main aspects of human activity that affect the distribution of woodlice. Neolithic farmers began the process of clearing primeval deciduous forests that dominated the landscape to create pastoral and later arable farmland. This opened up the landscape, enabling species associated with grassland, heathland and disturbed ground to become much more widespread than had been possible previously. Many species associated with human activity also occupy semi-natural habitats that are subject to disturbance as a result of natural processes. Examples include *Androniscus dentiger* and *Cylisticus convexus* associated with eroding coastal areas or *Haplophthalmus danicus* and *Trachelipus rathkii* along seasonally inundated river flood plains. These species are perhaps pre-adapted to colonise humanly disturbed (synanthropic) habitats. For some woodlice it is possible that the reverse is true. Heavily synanthropic species such as *E. caelatum* and *Buddelundiella cataractae* may have been introduced initially as synanthropes and may have colonised semi-natural 'eroding' coastal habitats subsequently.

The second relevant aspect is the movement of people and goods from continental Europe. Many of our species, including some of our more scarce woodlice, are probably ancient introductions unintentionally transported into Britain and/or Ireland by Mesolithic traders, Neolithic farmers or Roman invaders many millennia ago. It is not difficult to perceive woodlice being transported with plant material, soil, timber, stone or animal feed/bedding. Species such as *Porcellionides pruinosus*, *Porcellio dilatatus* and *P. laevis*, which have their distribution centred on the Mediterranean Basin (Schmalfuss, 2004), are likely examples. The once widespread practice of dumping solid ships' ballast at port is likely to be an under-rated source of species colonisation. To appreciate the effectiveness of this mechanism one only has to look at the large proportion of familiar British woodlice, including *T. rathkii* (Rapp, 1988), accidentally introduced from Europe and now dominating much of the North American isopod fauna.

This process of accidental introduction is ongoing. Our modern predominantly man-made landscape provides the opportunity for many species to spread considerably beyond their natural range. The effects of climatic extremes are ameliorated within our cities and industrial complexes (the 'urban heat island' effect) and this has allowed certain thermophilous synanthropes to survive much further north than would be possible otherwise. Some of our most recent arrivals include glasshouse aliens, such as *Styloniscus mauritiensis* and *Venezillo parvus*, undoubtedly brought in with tropical plant collections. Additional species may still be found in Britain and Ireland, particularly in glasshouses, nurseries and garden centres.

Other factors affecting distribution

Woodlice generally favour calcareous habitats. Some, such as the trichoniscid woodlice *Haplophthalmus montivagus* and *Trichoniscoides helveticus*, are restricted to limestone areas. Other calcicoles are able to exploit additional sources of calcium. *Armadillidium depressum* and *Porcellio spinicornis* may penetrate far from calcareous strata through their association with mortared walls, while *A. vulgare* and *A. nasatum* may inhabit non-calcareous localities on the coast where they utilise seawater spray and weathered sea-shells. *Platyarthrus hoffmannseggii* and *Trichoniscus provisorius* appear to be calcicoles, but it is more probable that these thermophilus species are exploiting the tendency of free draining limestone soils to warm up quickly. Only the ubiquitous *Oniscus asellus* and *Porcellio scaber* (and possibly *Trichoniscus pusillus* seg.) seem to be able to tolerant acidic conditions.

Ultimately, it is climatic factors, such as temperature and rainfall, that limit the distribution of a given species. Using these factors, the British and Irish isopod fauna can be broken down into three main biogeographic elements; Atlantic, continental and Mediterranean (Hopkin, 1987a; Sutton & Harding, 1989). It is apparent from the distribution maps that many species exhibit quite clear geographical distribution patterns across Britain and Ireland. An understanding of these three biogeographical elements helps explain these observed patterns.

Britain and Ireland experiences an Atlantic climate, characterised by a minimal difference (c. 10°C) between the relatively warm winter lows (January mean) and the relatively cool summer peaks (July mean). The prevailing south-westerly winds blowing over the Atlantic ocean bring a relatively high rainfall to Ireland and western Britain, typically more than twice as much as eastern regions. South-eastern parts of Britain, isolated from the temperature buffering effects of the Atlantic ocean, experience wider extremes of temperature. This is more akin to the continental climate of central Europe, where the summers are relatively hot, but the winters are cold enough to allow snow to settle on the ground. The southern fringes of Europe experience a Mediterranean climate where the summers tend to be hot and dry, but the mild rainy winters are similar to those of the Atlantic region. Similar conditions may prevail in the sunnier parts of southern England.

Atlantic species

It is species with an Atlantic distribution that best characterise our woodlice fauna. These have their European distribution centred on the oceanic fringe of western Spain, western France, the British Isles and north to Denmark and southern Scandinavia. Strict Atlantic species such as *O. asellus* ssp. *occidentalis*, *Oritoniscus flavus* and *Porcellionides cingendus* typically exhibit a distinct south-western distribution in Britain and Ireland. However, in the case of *Acaeroplastes melanurus* and *Eluma caelatum*, which have probably been introduced, this is not apparent. Three of our most widespread species, *O. asellus* ssp. *asellus*, *Philoscia muscorum* and *P. scaber*, have a broad Atlantic distribution and are abundant over much of western Europe, but become increasing scarce (and increasingly synanthropic) in central and eastern Europe.

Continental species

Continental species are characteristically distributed across central Europe, from north-eastern France to the Russian borders. Those that reach Britain are at the north-western limit of their global range. The central European species *L. hypnorum*, *H. montivagus* and *T. rathkii* are more or less confined to south-eastern England where the influence of the continental climate is most pronounced. Others, such as *Haplophthalmus danicus*, *Platyarthrus hoffmannseggii* and *Armadillidium vulgare*, are more widespread in Britain and Ireland, but still exhibit a predominantly south-eastern bias. It is probable that insufficiently warm summers (rather than prolonged cold winters) define the northern and western limits of their respective ranges. In western areas the relatively warm wet winters may put these continental species at a disadvantage.

The northern European species *Armadillidium pulchellum* and *A. pictum* demonstrate a north-western bias, being widely distributed in upland regions of north-western England, but rare or absent in the south-east. High summer temperatures (July mean) probably prevents them from penetrating south-eastern areas, but since neither penetrates the northern parts of Scotland then the northern limit of their distribution is probably determined by minimum winter temperatures (January mean). *Haplophthalmus mengii* seg. also appears to have a preference for north-western areas, first noted by Hopkin (1987a).

Mediterranean species

Mediterranean species, such as *A. depressum* and *A. nasatum*, typically have a southern bias to their British and Irish distributions. They are tolerant of our wet Atlantic winters, but require hot summers with high levels of insolation. Either insufficiently warm summers or prolonged cold winters may define their northern limit. It is apparent that *Porcellio dilatatus*, *P. laevis* and *Porcellionides pruinosus* are widely dispersed beyond southern England. Although originally Mediterranean, they have become widespread through their association with livestock (horses and stables were ubiquitous until the start of the 20th Century) and they are able to exploit the warmth generated in compost/manure heaps. Some current models of long-term climate change predict a shift in Britain's climate towards a more pronounced Mediterranean influence (Hadley Centre website, 2008) that may favour this suite of species.

Coastal species

A relatively high proportion of our woodlice fauna that is associated with the Atlantic and Mediterranean biogeographical zones is restricted to coastal areas. A few halophiles, *Ligia oceanica*, *Miktoniscus patiencei*, *Halophiloscia couchii*, *Stenophiloscia glarearum* and *Armadillidium album*, are typically found within a few metres of the extreme high tide mark. *Metatrichoniscoides celticus* and *Trichoniscoides saeroeensis* are rarely found more than a few kilometres inland. The majority of these species exhibit an Atlantic distribution (along the western coast of Europe), but *H. couchii*, *S. glarearum* and *A. album*, exhibit an extended Atlantic/Mediterranean distribution to include the northern coast of the Mediterranean Sea.

The distribution patterns of other species are also strongly influenced by maritime influences where the effects of latitude and/or acidic substrates are ameliorated. Species in other biogeographic groups, such as *P. hoffmannseggii* and *A. vulgare*, become increasingly coastal in northern areas where temperatures may be consistently a few degrees higher than adjacent inland areas.

Conservation

Through the efforts of the Non-marine Isopod Recording Scheme the terrestrial woodlice are among the best-studied invertebrate taxa in Britain and Ireland. The aquatic waterlice remain less well known. The scheme holds a substantial database containing up-to-date information on the distribution and habitat preferences of the British and Irish species. We have a thorough understanding of species taxonomy and a considerably improved understanding of the habitat requirements and distribution patterns of the more scarce members of our fauna.

Our fauna in a European perspective

Our fauna is mainly composed of widespread European species. There is a strong Atlantic element, with many species adapted to our relatively mild and damp climate. With the possible exception of *Metatrichoniscoides celticus*, we have no endemic species such as those found in isolated mountainous regions and glacial refuges of continental Europe. Species that arrived from continental Europe by genuinely natural means can be considered to be 'true natives'. These are likely to include species, such as *Ligidium hypnorum*, *Haplophthalmus montivagus*, *Trichoniscoides albidus*, *T. helveticus*, *Armadillidium pictum*, *A. pulchellum* and *Trachelipus rathkii*, that are associated with semi-natural habitats, including deciduous woodland, riverside meadows and calcareous grassland. The exclusively coastal species, such as M. *celticus*, *Miktoniscus patiencei*, *Trichoniscoides saeroeensis*, *Halophiloscia couchii* and *Armadillidium album* may also fall into this category.

It has become increasingly clear that some of our rarest species, possibly including *Buddelundiella cataractae*, *Eluma caelatum* and *Acaeroplastes melanurus*, may be ancient introductions unintentionally transported to our shores since the Mesolithic (Corbet, 1962). This should not detract from the conservation value of these species. Many of our rare and vulnerable invertebrates are believed to be ancient introductions and those that have successfully invaded semi-natural habitats are likely to have been overlooked as natives.

For example, the snail *Monacha cartusiana* is described by Kerney (1999) as "a 'weed' of cultivation" yet this species receives Red Data Book status. The indirect nature of the evidence used to identify these ancient introductions means that it is usually impossible to distinguish between native species and naturalised introductions. Thus, in the case of invertebrate conservation, including woodlice and waterlice, this traditional distinction is not a helpful consideration.

Conservation status

Comments on the conservation status detailed below are applicable to Britain only, since in Ireland rare and vulnerable species have not been allocated formal statuses.

UK Biodiversity Action Plan

The criteria for selection of species for inclusion on the UK Biodiversity Action Plan takes into account two main aspects. Firstly, whether the UK populations are of global importance and secondly, whether there are potential threats and/or observed declines in the UK. A review of UK BAP Priority Species, using IUCN criteria (IUCN, 2001) was published in 2007 (www.ukbap.org.uk/). All the non-marine isopods proposed for consideration were rejected as Priority Species status.

Red Data Book

Harding and Sutton (1985) gave provisional 'red data book' status to a number of species. This information was reviewed and updated by the British Red Data Book (Bratton, 1991), which lists our rare and vulnerable species. Two species, the stygobitic asellid *Proasellus cavaticus* and the oniscid *Armadillidium pictum* were assigned the status of RDB3 (Rare) and *Metatrichoniscoides celticus* was assigned RDBK (Insufficiently Known). Unlike Britain, Ireland does not have a Red Data Book listing for invertebrates. *Acaeroplastes melanurus* should be considered vulnerable, if not endangered, in Ireland. This species, which was considered extinct by Harding and Sutton (1985), has been rediscovered in Ireland, but surveys indicate that the extent of suitable habitat has declined considerably (Anderson, 2007). Other species, including *Miktoniscus patiencei* and *Halophiloscia couchii*, may also be vulnerable in Ireland.

Nationally Scarce species

Unlike many insect groups, a complete review of the conservation status of British woodlice and waterlice has never been formally published. However, the status of Nationally Scarce (previously known as Nationally Notable/Nb) has been assigned to a number of species and has been adopted by biological recording databases, such as Recorder, a standard source of reference to those working in conservation. Ten species of woodlice are assigned Nationally Scarce/Nb and the waterlouse *P. cavaticus* has been downgraded from RDB3 to Nationally Scarce/Nb. The list includes several species once considered very rare, such as *H. couchii* and *Eluma caelatum* (Harding & Sutton, 1985), which are now known to be much more widespread than previously thought. These species are listed in the Table 4 below. The Red Data Book and Nationally Scarce statuses of our entire asellid and oniscid fauna are in urgent need of updating. This revision should be undertaken through the application of IUCN criteria (IUCN, 2001). The use of data collated by the Non-marine Isopod Recording Scheme will be essential in completing this reassessment.

Table 4. Species of conservation value in Britain (not applicable to Ireland).

Scarce/Nb: Species thought to occur in 31 to100 10km squares of the National Grid, or for less well-recorded groups, between eight and twenty vice-counties.

RDB3: Species thought to occur in 15 or fewer 10km squares of the National Grid, or if more widely distributed, where populations occupy small areas of vulnerable habitat.

RDBK: Species that are suspected to belong to a RDB category, but because of lack of information this is not definitely known.

Species	Status	Principle habitat
Proasellus cavaticus	Scarce/Nb	Caves and wells
Buddelundiella cataractae	Scarce/Nb	Mainly coastal, but synanthropic sites inland
Haplophthalmus montivagus	Local *	Calcareous woodland and scrub
Metatrichoniscoides celticus	RDBK	Coastal, mainly supralittoral, erosion banks
Metatrichoniscoides leydigii	None *	Synanthropic habitats
Miktoniscus patiencei	Scarce/Nb	Coastal, supralittoral, saltmarsh, etc
Trichoniscoides albidus	Scarce/Nb	Wet woodland, scrub, alluvial meadows and coastal sites
Trichoniscoides helveticus	Scarce/Nb	Calcareous woodland, scrub and grassland
Trichoniscoides saeroeensis	Scarce/Nb	Coastal, mainly supralittoral, shingle, erosion banks, etc
Trichoniscoides sarsi	None *	Synanthropic habitats; churchyards and gardens
Halophiloscia couchii	Scarce/Nb	Coastal, supralittoral, rock and shingle
Stenophiloscia glarearum	Scarce/Nb	Coastal, supralittoral, shingle
Oniscus asellus occidentalis	None *	Rural habitats in south-western areas
Armadillidium pictum	RDB3	Upland rocky woodland, limestone pavement
Armadillidium album	Scarce/Nb	Coastal, mainly supralittoral, sand-dunes, etc
Armadillidium pulchellum	Scarce/Nb	Upland grassland, heathland, moorland and coastal sites
Eluma caelatum	Scarce/Nb	Mainly coastal, but synanthropic sites inland

* status needs revision

Species of conservation interest

The distribution map suggests that *Haplophthalmus montivagus* should be upgraded from 'local' to Nationally Scarce/Nb status. On current evidence *Metatrichoniscoides celticus* is endemic to the British Isles, but it is notoriously elusive, even at its known localities. It may have been overlooked on the adjacent coasts of north-western France and for this reason it has been rejected as a UK BAP Priority Species. The rare *Trichoniscoides sarsi*, currently known from several sites in eastern and central England and eastern Ireland, should be considered. On current evidence, *Metatrichoniscoides leydigii* is a casual introduction in Britain and of little conservation value. However, if populations occurring in coastal habitats, akin those reported in the Netherlands by Berg, Soesbergen, Tempelman and Wijnhoven (2008) then this species should also be considered.

Although it has proved to be more widely dispersed than previously thought, *A. pictum* remains one of Britain's rarest woodlice. It is exclusively associated with semi-natural habitats, typically unmanaged woodland, shady grassland or limestone pavement. It should be considered to be vulnerable to unsympathetic forest management, including clear felling or replanting with conifers, overgrazing and removal of limestone pavement. The Countryside Commission for Wales has recently commissioned a Common Standards Monitoring of this species in Wales (Joint Nature Conservation Committee, 2008).

The asellid *P. cavaticus* occurs as two distinct regional morphological forms in Britain. Proudlove, Wood, Harding, Horne, Gledhill and Knight (2003), speculate that if size variation in British populations is genetically linked then the distinct Mendip populations may be a separate cryptic taxon possibly endemic to the British Isles. If this were confirmed by genetic studies, this taxon may be of global importance and should be considered as a candidate UK BAP Priority Species. Ground-waters have been increasingly exploited and pollution of ground-waters may have become a problem in some areas. As with other subterranean aquatic crustaceans in Britain, *P, cavaticus* is likely to be susceptible to changes in ground-water quality (Proudlove *et al*, 2003).

The south-western taxon *O. asellus occidentalis* may be Britain's most vulnerable non-marine isopod (D.T. Bilton, personal communication). Considering that Britain provides a global stronghold for this taxon, then its conservation status should be set accordingly. In stark contrast, the more competitive *O. asellus asellus* is one of our most abundant species. The two taxa are genetically distinct and of ancient divergence (Bilton, Goode & Mallet, 1999). The association of *O. a. asellus* with human activities has allowed this expansive form to infiltrate south-western areas where originally *O. a. occidentalis* would have been the sole taxon (Bilton, 1994). Since the two taxa hybridise, the ultimate long-term fate of the less competitive *O. a. occidentalis* may be gradual hybridisation into extinction within Britain, if not globally.

It is apparent from Table 4 that the majority of the species listed are confined to coastal regions, including several associated with the supralittoral zone. *Miktoniscus patiencei*, *Stenophiloscia glarearum* and *Armadillidium album* remain infrequently recorded and, considering their preference for undisturbed coastal habitats, they should be considered potentially at risk throughout their British and Irish ranges. This is particularly true of *A. album*, which at some localities may be threatened by heavy tourist pressures, removal of strandline material and commercial sand extraction. However, recent records indicate that *H. couchii* and *Trichoniscoides saeroeensis* are fairly widespread within their respective ranges, whilst *Buddelundiella cataractae* and *E. caelatum* are not associated with the supralittoral zone and are able to utilise synanthropic sites inland. These latter four species are therefore less likely to be vulnerable to habitat loss and degradation than the three species considered above.

Of the remaining species listed in the table, *Haplophthalmus montivagus*, *Trichoniscoides albidus* and *Trichoniscoides helveticus* (as with *A. pictum*) are mainly associated with semi-natural woodland/scrub and upland calcareous grassland/limestone pavement. *Armadillidium pulchellum* also inhabits moorland and heathland. Britain and Ireland, together, hold a significant proportion of the global population of *A. pulchellum* (Harding & Sutton, 1985), which is relatively rare elsewhere in Europe.

Conservation of woodlice and their habitats

Our woodlice species and their habitats are being increasingly put under threat by our modern way of life. Intensification of agricultural and forestry practices, the 'development' of brownfield sites and the stabilisation our coastline are all likely to affect woodlice populations. In addition, species distribution patterns seen today are not static and in light of predicted long-term climatic change it is very likely that many species will experience an expansion or contraction of their ranges over coming decades. Species associated with semi-natural habitats, such as *Armadillidium pictum* and *A. pulchellum*, are likely to be poor dispersers and potentially vulnerable to climate change. Conservation, therefore, should not be a matter of protecting rare species in isolated refuges, but should be approached from a landscape perspective.

In common with many invertebrates, the diversity of woodlice found at a given site does not necessarily correlate with botanical richness. The most important factors tend to be microclimate, and, in the case of soil-dwelling species, soil structure. In many cases, long-term continuity of habitat is essential. Some of our rarest woodlice occupy specialist habitats that are often perceived by the general public to be of little value for conservation. These include veteran trees (e.g. *A. pictum*), wet woodland (e.g. *Trichoniscoides albidus*), slumping cliffs (e.g. *Eluma caelatum*), shingle beaches (e.g. *Stenophiloscia glarearum*), erosion banks (e.g. *Metatrichoniscoides celticus*) and brownfield sites (e.g. *E. caelatum*).

Key woodlouse habitats

Woodlice may be found in almost anywhere, but certain habitat types frequently support rare species or assemblages of our less common species. The UK Biodiversity Action Plan (www.ukbap.org.uk/) includes 'Broad Habitat Statements' that provide summary descriptions of 28 natural, semi-natural and urban habitats. Within this Broad Habitat classification there are listed a larger number of UK BAP Priority Habitats, each with a detailed action plan for conserving these habitats. Those habitats that are particularly relevant to terrestrial oniscids are as follows:

Broad Habitats
  Built up areas and gardens
  Urban

UK BAP Priority Habitats
  Coastal saltmarsh
  Coastal sand dunes
  Coastal vegetated shingle
  Maritime cliff and slopes
  Upland oakwood
  Upland mixed ashwoods
  Wet woodland
  Limestone pavements
  Upland calcareous grassland
  Lowland meadows

Coastal habitats

Coastal habitats, such as shingle beaches, erosion banks, sand dunes and soft cliffs, generally support the most diverse woodlice assemblages. They are slow to establish and are entirely reliant on the natural processes of erosion and deposition for their existence. In their natural state they do not require management to maintain their wildlife interest. The main threats to coastal habitats are coastal stabilisation, for example through engineered sea defences, the extraction of sand or shingle and offshore dredging. These alter the natural dynamic processes upon which coastal habitats depend. Increased demand for recreation can cause damage to the supralittoral strandline, for example through mechanical beach cleaning, the increased use of off-road vehicles or the removal of driftwood. Pollution, especially by oil, can be damaging if washed into the supralittoral zone. An increasing threat is long-term climate change. Predicted increased severity of storms and rising sea levels are likely to affect our coastal fauna, both directly and indirectly through the perceived need for additional coastal stabilisation.

Synanthropic habitats

Synanthropic habitats, that have been substantially altered by human activity, share a number of features with coastal habitats, such as low fertility and sparsely vegetated structure, and are arguably the next most significant habitat for woodlice. They can support many Nationally Scarce species, otherwise found in naturally disturbed coastal habitats, that are able to exploit humanly disturbed sites such as churchyards, allotments, farmyards or garden centres. One important group of synanthropic habitats are 'brownfield sites' such as waste ground in towns, disused quarries and pits, old railway lines, disused airfields and post-industrial sites. Aided by government policy, many brownfield sites are being lost to urban sprawl and to industrial development. Another threat is that of well-meaning local conservation groups who through the importation of soil and by planting wildflowers or trees may eradicate the actual conservation value of a brownfield site.

Upland woodlands

Upland oakwoods tend to occur on more acid substrates whereas ash favours base-rich soils. Upland woodlands are an important habitat for the Red Data Book *Armadillidium pictum*, and also the widespread *H. mengii*. *A. pictum* will use dead wood habitats which are typically found in unmanaged woodland. During the last century many woodlands have been degraded through unsympathetic management, such as clear felling and/or replanting with conifers. Upland woodlands in particular are vulnerable to overgrazing, which prevents natural regeneration and affects species composition.

Lowland wet woodlands

Lowland wet deciduous woodland, such as found on river floodplains or on fens, is a rare habitat in Britain. Associated woodlice are *Ligidium hypnorum*, *Trichoniscoides albidus* and *Haplophthalmus montivagus*. Fallen dead wood and the development of a deep litter layer will add to the diversity of the woodlice fauna. As with upland woodland there have been considerable losses of wet woodland habitats, mainly due to replanting with conifers, but also through conversion to agriculture. Wet woodlands are dependent on hydrology. Lowering of water-tables through drainage or groundwater abstraction and pollution of groundwater or nutrient enrichment from agricultural run-off are detrimental to this habitat

Upland calcareous grasslands

Upland calcareous grasslands and limestone pavement support a small but specialist woodlice community. Short-turf or sparsely vegetated grasslands often support *Armadillidium pulchellum*, while in shady grasslands the rare *A. pictum* may also be present. One of the principal threats to limestone pavement is its illegal removal for use in garden rockeries. Another threat affecting pavements and grasslands alike is overgrazing or undergrazing with livestock, which affects the structural diversity of the swards. Limestone quarrying may cause local removal of grasslands, but equally damaging is the in-filling of abandoned and re-vegetated quarries.

Lowland neutral meadows

In southern England, lowland neutral meadows and pasture that are seasonally inundated support a distinctive woodlice community, characterised by *L. hypnorum*, *Haplophthalmus mengii*, *T. albidus* and *Trachelipus rathkii*. Lowland meadows often form part of a larger habitat mosaic that includes disused gravel pits, which further enhances the species diversity. In the last century most lowland meadows were lost through agricultural improvement by the application of fertilisers, re-seeding or conversion to arable. Many sites are typically dependent on high water tables and/or seasonal inundation. Land drainage, flood defences, ground water abstraction and other activities reduce the frequency and duration of flood events and therefore adversely affect species composition.

Collecting and recording

Collecting waterlice

Asellids inhabit a wide variety of slow moving or still water-bodies, including ponds, lakes, ditches, canals, sluggish rivers and marshes. Specimens may be collected by hand where the water is shallow and accessible from the shore. This method can be productive where there are large stones or dead wood that can be rolled over to reveal specimens beneath. However, waterlice are most successfully collected with a pond net and using a white tray to sort the samples. Specimens can be obtained amongst mud, dead leaves or other rotting debris by disturbing the top few centimetres of the substrate with a net and sweeping through the resultant cloud of debris. Where a gravely or stony substrate occurs this too can be disturbed by stirring or wading, prior to netting. Where there is some movement of water the net should always be placed downstream of the disturbed substrate. Where there are dense stands of aquatic or marginal vegetation, including reed beds, the net should be worked vigorously through the vegetation. *Proasellus cavaticus* is entirely subterranean and, since it is very rarely seen in surface water issuing from springs, specialist knowledge of speleology is useful. Although usually collected from shallow pools or water films in caves, it has also been collected from wells and very rarely from places where ground water issues to the surface.

Collecting woodlice

An introduction to collecting methods is given in Lee, Barber and Gregory (2007). Four species, *Trichoniscus pusillus*, *Philoscia muscorum*, *Oniscus asellus* and *Porcellio scaber* are easily found and are often abundant. In the south and east *Armadillidium vulgare* is also common. Other species are much less frequently encountered, but locally some may be widespread and relatively common once their precise niche is discovered. In order to record a wide array of species it is essential to examine a wide variety of habitats and microsites. If the same habitats and microsites are always searched it is quite likely that the same species will always be found. However, the discovery of some of our most elusive species can depend on good luck; simply being in the right place at the right time. The best time to collect woodlice is during spring or autumn, or even during the depths of winter. Woodlice can be collected in summer, but they often become elusive during dry weather.

There are a variety of ways of collecting woodlice and each will produce different results. The simplest method is to turn over stones and logs and to examine the underside of the turned object carefully for small species. Careful hand sorting of substrates, such as friable soil, rotten wood, leaf litter, moss, rubble, compost heaps, coastal shingle or strandline debris can reveal many elusive species. Alternately, the material can be sieved onto a pale blue sheet (white will suffice). Pitfall traps will typically catch large numbers of common species, but in some habitats, for example coastal shingle, some very rare species such as *S. glarearum* could be caught. Woodlice are nocturnal and those species that inhabit inaccessible crevices on walls, sea cliffs or shingle beaches can be much easier to survey at night by torchlight. This is a good method for recording *Porcellio spinicornis* and *Armadillidium depressum* in built-up areas. It is through the use of these more unusual techniques that our understanding of the biology and distribution of some of our lesser-known species will improve.

Trichoniscid woodlice

The 'pygmy woodlice' belonging to the family Trichoniscidae are mainly soil-dwelling and are rarely encountered by causally turning over stones or logs. For these small, elusive and generally less conspicuous species recorder bias remains an ever-present problem. A comparison of the distribution map of *Oniscus asellus*, a large and conspicuous woodlouse, with that of *Trichoniscus pygmaeus*, a small soil-dwelling trichoniscid that is probably just as widespread, clearly illustrates this point. Thus, it is necessary to adopt a targeted approach to successfully record trichoniscids. It is essential to search in the appropriate microsites, to collect in the appropriate seasons and to recognise specimens in the field that appear to be different species.

The more elusive trichoniscids are soil-dwelling species generally associated with friable soils, often where some organic (peaty) material has accumulated. The easiest method to locate them is carefully examine the underside of large stones or logs partly embedded in the soil surface. To find really elusive species, such as *Trichoniscoides* spp., it is often necessary to remove and examine carefully rocks or rubble buried beneath larger rocks. Searching in a damp spot, whether this is near a watercourse, flushed areas or within damp hollows, is likely to be most productive. Some species may also occur within rotten wood (e.g. *Haplophthalmus montivagus*), deep leaf litter (e.g. *Trichoniscoides albidus*) or well-rotted compost heaps (e.g. *Haplophthalmus danicus*).

Trichoniscids are most easily found during periods of damp mild weather in autumn and spring. It is usual to find small numbers of other species among large numbers of the ubiquitous *Trichoniscus pusillus* agg., a species that rapidly seeks shelter when disturbed. Thus, look out for specimens that are reluctant to move even when provoked as these may prove on close examination to be something different, such as *T. pygmaeus* or even *Trichoniscoides* spp.. Perhaps more surprising is the well-documented occurrence of the notoriously elusive *Trichoniscoides* spp. during heavy midwinter frosts, even beneath even small pieces of surface debris (Doogue & Harding, 1982; Harding & Sutton, 1985; Hopkin, 1991a; Daws, 1995b). It is possible to find these species in mid summer, but this typically entails time consuming and tedious hand searching of soil up to half a metre in depth. Turning over a few stones in winter is much more efficient.

Identification keys

The study of woodlice has progressed considerably since the days of Edney (1953, 1954), whose identification keys rely on characters that are difficult to see and are difficult for a beginner to use. A good general introduction to the biology of woodlice can be found in Stephen Sutton's book *Woodlice* (Sutton, 1972), which includes a number of useful line drawings and a series of colour illustrations of selected species. The identification key, which is also published independently as Sutton, Harding and Burn (1972), is now considerably out-of-date since several additional species are now known to occur in Britain. Harding (1990) lists all literature relevant to the biogeography and ecology of British woodlice published from 1830 until 1986 (other than the publications of W.E. Collinge, which are listed in Harding, 1977). Doogue and Harding (1982) list all publications (1836-1981) relevant to Ireland.

Identification of all our native and naturalised species is covered by the AIDGAP key by Steve Hopkin (Hopkin, 1991a), which includes colour plates and a clear illustrated key specifically aimed at beginners. The Synopsis by Oliver and Meechan (1993) gives a much more detailed account of each species, including line drawings of male genitalia and secondary sexual characters of all species. The most accessible key to all four species of British Asellidae is the Freshwater Biological Association key by Gledhill, Sutcliffe and Williams (1993), which includes other species of Malacostraca Crustacean. Additional information can also be found in the newsletter and the Bulletin of the British Myriapod and Isopod Group (BMIG).

Critical species

Some common species may be misidentified as scarce species to which they bear a superficial resemblance, either because of unusual colour or size variation, problems with using identification keys or sometimes simply over-enthusiastic recording. The reverse is also true and care should be taken in the field to avoid overlooking scarce species among the background of more common ones. The species that most frequently cause problems are listed in Table 5 below. In the case of a few critical species (e.g. *Trichoniscoides* spp.) a male specimen must be collected for reliable identification and a voucher specimen should be retained. These critical species are listed in Table 6.

Table 5. Frequently confused waterlice and woodlice.

More widespread species	Scarce species	More widespread species	Scarce species
Asellus aquaticus	*Proasellus meridianus*	*Armadillidium vulgare*	*Armadillidium album*
Ligia oceanica juveniles	*Halophiloscia couchii*	*Armadillidium vulgare*	*Armadillidium depressum*
Haplophthalmus danicus	*H. mengii* or *H. montivagus*	*Armadillidium vulgare*	*Armadillidium nasatum*
Trichoniscoides sp.	*Metatrichoniscoides* sp.	*Armadillidium vulgare*	*Eluma caelatum*
Trichoniscus pusillus	*Oritoniscus flavus*	*Armadillidium pulchellum* or *A. vulgare*	*Armadillidium pictum*
Trichoniscus pusillus	*Trichoniscoides albidus*		
Trichoniscus pusillus juveniles	*Trichoniscus pygmaeus*	*Armadillidium pictum* or *A. vulgare* juveniles	*Armadillidium pulchellum*
Trichoniscus pygmaeus	*Trichoniscoides* sp.		
Philoscia muscorum	*Ligidium hypnorum*	*Porcellio scaber*	*Porcellio dilatatus*
Philoscia muscorum	*Porcellionides cingendus*	*Porcellio scaber*	*Porcellio laevis*
Oniscus asellus or *Porcellio scaber*	*Porcellio spinicornis*	*Porcellio scaber* or *Oniscus asellus*	*Trachelipus rathkii*

Table 6. Critical species pairs requiring a male specimen to be certain of identity.

More widespread species	Scarce species	More widespread species	Scarce species
unpigmented *Asellus aquaticus*	*Proasellus cavaticus*	*Trichoniscoides saeroeensis*	*Metatrichoniscoides celticus*
Haplophthalmus mengii seg.	*Haplophthalmus montivagus*	*Trichoniscoides saeroeensis*	*Trichoniscoides sarsi*
Metatrichoniscoides celticus	*Metatrichoniscoides leydigii*	*Trichoniscoides sarsi*	*Trichoniscoides helveticus*

Submitting records

An updated non-marine isopod recording card (RA84) has recently been produced (see Figure 48) and is available from BRC. This should be used to record field observations of a single visit to a single site. Alternately records may be submitted electronically (see section below). Information provided with the record should be as detailed and as complete as possible. A species record consists of a minimum of four pieces of information.

- What species was found (species name)
- Where was the specimen found (locality name and grid reference)
- When was the specimen found (date)
- Who found/identified the specimen (collector name and/or determiner name)

Species

It is essential that all species found on a site are recorded, even the common ones. Records of 'rare' and 'interesting' species are much more useful if details of other associated species are listed. If the identification is based on a male specimen this should be noted. For critical species pairs, such *Haplophthalmus mengii* and *H. montivagus*, *Trichoniscoides sarsi* and *T. helveticus* or *Trichoniscus provisorius* and *T. pusillus* seg., this is essential, otherwise the record will be attributed to the less precise species aggregate.

Site

Species records should be site-based. Where possible a site name that occurs on the Ordnance Survey 1:50,000 series map should be used. A separate species list should be compiled for each site visited, ideally a different list for each microsite examined (e.g. under stones, amongst leaf-litter, etc). Species lists compiled for a tetrad (2km squares) or 10km squares are of little value and should be avoided.

Grid Reference

If the sample is collected from a clearly defined area a six-figure (100m) OS grid reference (e.g. SP123456), or if using a GPS an eight-figure (10m) grid reference, should be used. A ten-figure (1m) GPS reading is false accuracy and should be avoided. If collecting was undertaken over a wide area then a four-figure (1km) grid reference (e.g. SP12-45-) may be more relevant. A vague 10km square grid reference (e.g. SP14) should be avoided.

Date

Where possible give the full date, including day, month and year. If trapping for a short period (e.g. Pitfall trap) give the start date and the end date. Vague dates, such as a single year or a year range (e.g. 1995 to 2001), should be avoided.

Recorder

It is essential to give the collector's name. If another person identified the specimen then their name should be included too. This is essential in the case of scarce species or species recorded in unusual circumstances if the record is to be accepted by the recording scheme.

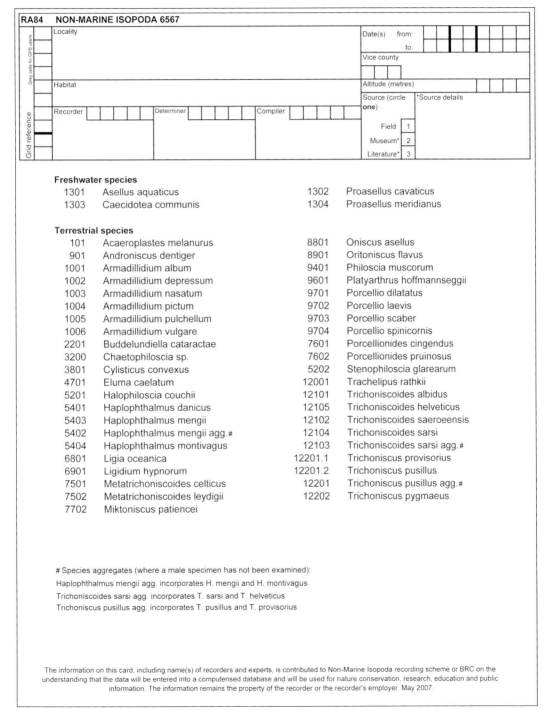

Figure 48. The revised RA84 non-marine isopod recording card (front).

PRINCIPAL HABITATS
Tick one only

MARITIME
- 010 rock
- 011 shingle
- 012 sand
- 013 mud/salt
- 014 saltmarsh
- 015 artificial construction
- 016 sea cliff

SAND DUNE
- 020 bare sand
- 021 tussocky
- 022 dense sward
- 023 dune slack
- 024 dune heath
- 025 machair

WETLAND
- 030 fen
- 031 carr
- 032 bog

HEATH/MOOR mainly
- 040 moss/lichen
- 041 grass/sedge/rush
- 042 heather
- 043 Vaccinium
- 044 bracken
- 045 gorse
- 046 mixed
- 048 other (specify)

GRASSLAND
- 050 ungrazed
- 051 lightly grazed
- 052 heavily grazed
- 053 mown

VARIOUS
- 060 park/recreation
- 061 orchard
- 062 tree nursery
- 063 young plantation
- 064 hop field
- 065 soft fruit field
- 066 vineyard
- 067 churchyard
- 068 urban/suburban open space

CULTIVATION
- 070 grass ley
- 071 cereal crops
- 072 other seed crops
- 073 root crops
- 074 Brassica/other veg
- 075 flower crops
- 076 garden, ornamental
- 077 garden, domestic

SCRUBLAND
- 080 dense
- 081 open with grass/herbs

WOODLAND mainly
- 090 mixed deciduous
- 091 poplar/aspen
- 092 birch
- 093 alder
- 094 hornbeam
- 095 hazel
- 096 beech
- 097 chestnut
- 098 oak
- 099 elm
- 101 acacia
- 102 field maple
- 103 sycamore
- 104 lime
- 105 ash

mixed coniferous
- 110 fir
- 111 spruce
- 112 pine
- 113 larch
- 114 yew
- 120 mixed coniferous/deciduous
- 130 other (specify)

BUILDING
- 140 outside
- 141 ruin
- 142 inside, uninhabited
- 143 inside, inhabited
- 144 glasshouse, unheated
- 145 glasshouse, heated

ROCK
- 150 flat/pavement
- 151 scree
- 152 inland cliff

CAVE/TUNNEL/WELL
- 160 threshold
- 161 dark zone

EXCAVATION
- 170 sand pit
- 171 gravel pit
- 172 opencast mine
- 173 quarry floor
- 174 quarry face
- 175 scree

WASTE GROUND
- 180 demolition site
- 181 spoil tip (specify type)
- 182 organic/refuse tip

OTHER (specify)
- 190

HABITAT DETAILS
Tick one in each section

- 001 Intertidal
- 002 Around High Water Mark
- 003 Splash zone
- 004 Coastal (<1km from sea)
- 005 Coastal (1-10km from sea)
- 006 Inland (>10km from sea)

- 001 Urban
- 002 Suburban/village
- 003 Rural

ECOTONE
- 010 Waterside — pool
- 011 lake
- 012 ditch/stream
- 013 river
- 014 canal
- 015 saline
- 020 Road/railside verge
- 021 Path and verge
- 030 Woodland edge/ride
- 040 Tree row
- 041 Hedge
- 050 Wall, with mortar
- 051 Wall, drystone
- 052 Wood fence
- 060 Other (specify)

MICROSITE LOCATION
- 001 Under
- 002 On
- 003 In

SUBSTRATE
- 001 Rock
- 002 Stone
- 003 Shingle
- 004 Soil/sand
- 005 Leaf/twig litter
- 006 Tussocks/clumps
- 007 Bark surface
- 008 Dead wood
- 009 Dead wood/under bark
- 010 Moss/lichen
- 011 Dung heaps
- 012 Carrion
- 013 Fungi
- 014 Nest (specify)
- 015 Shore line jetsam
- 016 Human rubbish/garbage
- 017 Other (specify)

SLOPE
- 001 Flat
- 002 Shallow (5-15°)
- 003 Definite (15-25°/1 in 4)
- 004 Steep (25-45°/1 in 2)
- 005 Very Steep (>45°/1 in 1)

ASPECT
- 001 Flat
- 002 Facing N
- 003 Facing E
- 004 Facing S
- 005 Facing W

SOIL
- 001 Heavy clay
- 002 Clayey
- 003 Peat
- 004 Loam
- 005 Sandy
- 006 Pure sand
- 011 Shallow (<10cm)

SOIL/ROCK
- 001 Calcareous
- 002 Non-calcareous

METHOD OF COLLECTION
- 001 Casual/turnover
- 002 Dig/Sieve/Sort
- 003 Pitfall trap
- 004 Extraction
- 005 Existing collection

COMMENTS eg
- Local topography (eg top/bottom of slope)
- Soil moisture status
- Litter type
- Behaviour
- Food
- Predators parasites
- Etc
- Site conservation status

Figure 48 (continued). The revised RA84 non-marine isopod recording card (reverse).

Additional information

Additional information, such as Watsonian vice-county (Dandy, 1969) and altitude, are also useful.

Habitat information

The reverse of the revised RA84 recording card (Fig. 48) includes habitat classification information. The principle habitat type being sampled is chosen from the left-hand side. Detailed habitat information, such as microsite, aspect and soil type, is selected from the options on the right-hand side.

Electronic records

It is becoming increasing useful to have species records submitted in an electronic format, whether as standard spreadsheets or via specialist biological recording packages. Data can be dealt with in a variety of formats, but if standardised as far as possible this will considerably reduce the time taken to process records submitted to the recording scheme. Ideally each species record, including the species name and all associated data, should comprise one row of a spreadsheet.

Date format

The format dd/mm/yyyy is preferred, e.g. 13/03/2002. If the day or month is unknown substitute '00', e.g. 00/03/2002 (March 2002) or 00/00/2002 (sometime in 2002). However, care should be taken with MS Excel as this can treat the '/' character as a division function and will corrupt the date information unless the worksheet is formatted as text. For date ranges (e.g. for pitfall traps) separate start date and end date fields (columns) should be used.

Grid reference format

The format SP123456, where the two letters denote the 100km square, or its numeric equivalent, i.e. 42/123456 is preferred. In the latter case use '/' to separate the 100km square from the main grid reference. If the full six-figure reference is not known use '-' to fill the gaps i.e. SP12-45- (1km reference) or SP1-4- (10km reference). Empty spaces should not be inserted between letters and/or numbers.

Source of record

The majority of records are made in the 'field'. However, it is necessary to differentiate between those records made in the field from those gleaned by examination of preserved museum specimens ('museum') or extracted from published literature ('literature').

Species aggregates

If you have examined a male specimen to confirm the identity of a specimen please add a comment to the effect of 'males seen'. For critical species pairs such *H. mengii/H. montivagus, T. sarsi/T. helveticus* or *T. provisorius/T. pusillus* this is essential, otherwise the record and its associated habitat data will be included within the broad species aggregate.

Habitat information

Ecological information associated with the species record is extremely valuable and appropriate fields (columns) should be compatible with those found on the reverse of the revised RA84 recording card (Fig. 48). This should include a principal habitat and more detailed habitat information, such as coastal/inland, urban/rural, ecotone, microsite, substrate, slope, aspect, soil, rock and comments.

Future recording

Distribution patterns are not static. The distribution maps presented in this work are a snap-shot of our current knowledge and cannot be considered to be a definitive statement. There remains much to be discovered about the waterlice and woodlice fauna of Britain and Ireland.

Many areas of our countryside remain poorly known. Scotland and Ireland have always been relatively under-recorded. Little is known of the asellid fauna of these countries and many of the oniscid records are over 25 years old. However, the efforts of the few active recorders indicate that there remains much of interest to be discovered in these areas. For example, recent survey work in Aberdeenshire has revealed many species considerably further north than previously known in Britain.

Many species are relatively under-recorded. The small soil-dwelling trichoniscid woodlice notoriously so, but also species such as *Stenophiloscia glarearum*, and possibly *Armadillidium pictum*. The situation is further compounded by the difficulty of identifying critical species pairs, such as *Haplophthalmus mengii*/*H. montivagus* or *Trichoniscoides sarsi*/*T. helveticus*. The latter species pair is of particular interest since the two species seem to have mutually exclusive distributions across England. Another suite of under-recorded species is that associated with farmyards. This includes *Porcellio dilatatus*, *P. laevis* and *Porcellionides pruinosus*. The fauna of synanthropic habitats is often diverse and always worth recording.

Many factors, such as chance introductions and long-term climatic change, will continue to alter the observed ranges of many species. Some species, including *Buddelundiella cataractae*, *Armadillidium depressum* and *Eluma caelatum*, are able to disperse passively in association with human activity. Many species may spread further north, but a few, such as *A. pictum*, may undergo a contraction of range as a result of predicted climatic change.

Most of the species mentioned above are rarely discovered by casual surveys. Thus, it is essential that appropriate sampling methodologies are applied to appropriate habitats in order to record these interesting species. Further survey work will help us understand the distribution of these under-recorded species and to determine what their habitat requirements actually are. Future recording will also provide data that will allow us to follow any changes in species distribution patterns that may be occurring. Considering the enthusiasm of the dedicated waterlice and woodlice recorders found throughout Britain and Ireland there is every reason to believe that this latest phase of recording will be as successful as previous ones.

Glossary

The following is not intended to be a comprehensive glossary of terms used in this volume but rather to explain some possibly unfamiliar words.

autochthonous Originating from within a given habitat, community or system.

diploid Having the normal double set of homologous chromosomes.

ecophysiological Pertaining to the physiological adaptations of organisms to their habitat or environment.

endemic A species native to, and restricted to, a particular geographic region.

endopod The inner branch of the crustacean limb. In woodlice, only the limbs of the pleon retain the primitive branched state. The endopods are flattened plates on the ventral surface except for the last pair which form the inner branch of the uropods.

epimera The projecting lateral edges of the dorsal plates (pereonites) of the body.

eurytopic Living in a wide range of habitats.

exopod The outer branch of the crustacean limb (see endopod). The last pair form the outer branch of the uropods.

flowstone A feature in caves where layers of 'stone' are built up through the deposition of dissolved calcite, or other carbonate minerals, where films of water flow along floors or sloping walls.

halophile A species thriving in saline habitats.

haplophthalmoid Having the characteristic dorsal sculpturing of the genus *Haplophthalmus* (e.g *Buddelundiella*).

hybrid swarm A series of highly variable forms produced by the crossing and back-crossing of hybrids (as found in *Oniscus asellus*).

hygrophilous Thriving in moist habitats.

hyphomycetes A group of fungi within the division Ascomycota. They are common on organic debris of all types whether terrestrial or aquatic. Some species, such as *Beauveria bassiana*, are pathogenic to arthropods.

interstitial Occurring within the pore spaces between soil or sediment particles.

isostatic Pertaining to the vertical movement of the land and sea due to uplift caused by melting of the ice-sheets (and therefore removal of considerable load) since the last Ice Age.

littoral Pertaining to the inter-tidal zone of the sea shore.

mesocavernous Pertaining to subterranean cracks and voids that occur throughout bed-rock, such as limestones.

microsite As used in the Non-marine Isopod Recording Scheme, the exact site where an animal was found (cf microhabitat: the place where an animal is usually found).

myrmecophilous Living in association with ants.

native (indigenous) Species that have originated in a given area without human involvement or that have arrived there without intentional or unintentional intervention of humans from an area in which they are native.

naturalised An alien or introduced species that has become established.

ocellus (plural ocelli) A simple light detecting structure composed of a single, typically pigmented, lens. Each eye, located on either side of the head, may be composed of one, three or many (more than eight) ocelli (or absent in blind species).

palaearctic Geographic region comprising Europe, north Africa, western Asia, Siberia, northern China and Japan.

parthenogenetic Refers to a species in which offspring are produced by a female that has not been fertilised by a male (e.g. *T. pusillus* seg.).

pereon The middle section of the body; of 7 segments, each with a pair of walking limbs (pereopods).

pereonites The dorsal plates of the pereon, one per segment.

pereopod The leg of a woodlouse or waterlouse.

phenology Study of natural phenomena, including species activity, in relation to weather and climate.

phreatic Pertaining to groundwater.

phyletic Pertaining to a line of direct descent, or a course of evolution.

pleon The rear section of the body. It is much shorter than the pereon and ends in the telson. The limbs are flattened plates modified for respiration (and further modified for sperm transfer in the male).

pleonites The dorsal plates of the pleon.

pleopodal lungs Air-filled tubes ramifying through the interior of the pleopod exopods, bathed in blood and have a respiratory function.

red-rot A late stage of wood decomposition where small chunks of decomposing wood (usually from a fallen tree) have turned red in colour.

scutellum The central dorsal ridge forming the 'forehead' of woodlice, particularly noticeable as a distinctive 'snout' (e.g. *A. nasatum*).

sensu lato In the broad sense. When referring to a taxon, used to indicate that other closely related taxa that are considered as distinct are included, i.e. within a species aggregate.

sensu stricto In the strict sense. When referring to a taxon, used to indicate that only the nominate taxon is included (and therefore excluding other taxa within a species aggregate).

species aggregate (agg.) A group of closely related species that are difficult to distinguish. For example, there are a number of woodlice species that can only be separated by microscopic examination of male specimen using either sexual characters (e.g. exopods or endopods) or secondary sexual characters (pereopods).

stygobite A species entirely associated with subterranean aquatic habitats (e.g. *P. cavaticus*).

supralittoral Pertaining to the region of sea shore immediately above the high water mark and subject to spray or wave splash (the splash zone).

synanthropic Living in association with man. Hence synanthrope.

synonym One of two or more scientific names applied to the same taxon. The name published first usually takes precedence. Hence, synonymous.

talus Rock debris accumulated at the base of a cliff or slope.

telson The posterior projection of the pleon. It is often pointed, but may be rounded or quadrangular.

tergite A dorsal plate (sclerite) of the cuticle. Hence tergal.

thermophilus Thriving in warm environmental conditions. Hence thermophilic.

triploid Having three sets of homologous chromosomes (cf diploid).

troglobite A species entirely associated with subterranean habitats.

troglophile A species able to thrive in subterranean habitats, but equally able to thrive above ground (e.g. A. *dentiger*).

uropods The much modified last pair of pleopods, typically protruding beyond the rear of the pleon, with 2 branches each side flanking the telson (but truncated and flush in Armadillidiidae).

vadose Pertaining to water in the Earth's crust above the level of permanent ground water.

xerophilous Thriving in dry habitats. Hence xerophilic.

References

Alexander, K.N.A. (1995). *Armadillidium pictum* Brandt (Isopoda: Armadillidiidae) new to Gloucestershire. *British Journal of Entomology and Natural History*, **8**: 76.

Alexander, K.N.A. (2000). A relict population of *Armadillidium pulchellum* (Zencker) (Isopoda: Armadillidiidae) in the heathlands of south-east England. *British Journal of Entomology and Natural History*, **13**: 133.

Alexander, K.N.A. (2008). *Platyarthrus hoffmannseggii* Brandt in arboreal ant nests (Isopoda, Oniscidea, Platyarthridae). *Bulletin of the British Myriapod and Isopod Group*, **23**: 15-16.

Anderson, R. (2007). Observations on the status and ecology of *Acaeroplastes melanurus* (Budde-Lund) (Crustacea: Oniscoidea) at Howth Head, Dublin. *Irish Naturalists' Journal*, **28**: 497-505.

Aston, R.J. & Milner, A.G.P. (1980). A comparison of populations of the isopod *Asellus aquaticus* above and below power stations in organically polluted reaches of the River Trent. *Freshwater Biology*, **10**: 1-14.

Barber, A.D. & Jones, R.E. (1996). Geographical distribution of diplopods in Great Britain and Ireland; possible causal factors. In: Geoffrey, J-J, Mauries, J-P & Nguyen Duy-Jaquemin, M. (eds), *Acta Myriapodologica. Mémoires Muséum national d'Histoire naturelle*, **169**: 243-256.

Barrett K.E.J. (1977). *Provisional Atlas of the Insects of the British Isles, Part 5, Hymenoptera: Formicidae*. Huntingdon, Natural Environment Research Council.

Berg, M.P. (2008). Distribution and ecology of two enigmatic species, *Trichoniscoides sarsi* Patience, 1908 and *T. helveticus* (Carl, 1908) (Crustacea, Isopoda) in the Netherlands. *Bu alletin of the British Myriapod and Isopod Group*, **23**: 2-8.

Berg, M.P., Soesbergen, M., Tempelman, D. & Wijnhoven, H. (2008). *Verspreidingsatlas pissebedden, duizendpoten en miljoenpoten (Isopoda, Chilopoda, Diplopoda)*. EIS-Nerderland, Leiden.

Bilton, D.T. (1990). Two *Oniscus* in Britain! *British Isopod Study Group Newsletter*, **No. 28**: 3.

Bilton, D.T. (1994). Intraspecific variation in the terrestrial isopod *Oniscus asellus* L. *Zoological Journal of the Linnean Society*, **110**: 325-354.

Bilton, D.T. (1995). Finds at the Cumbrian meeting, 1995. *British Isopod Study Group Newsletter*, **No. 38**: 3.

Bilton, D.T., Goode, D. & Mallet, J. (1999). Genetic differentiation and natural hybridization between two morphological forms of the common woodlouse, *Oniscus asellus* Linnaeus, 1758. *Heredity*, **82**: 462-469.

Bouchon, D., Rigaud, T. & Juchault, P. (1998). Evidence for widespread Wolbachia infection in isopod crustaceans: molecular identification and host feminization. *Biological Sciences*, **265**: 1081-1090.

Bratton, J.H. (1991). *British Red Data Books: 3. Invertebrates other than Insects*. Joint Nature Conservation Committee. Peterborough.

Cawley, M. (1993). Upland populations of *Trichoniscoides saeroeensis* Lohmander in North West Ireland. *British Isopod Study Group Newsletter*, **No. 36**: 3.

Cawley, M. (1996). The woodlice (Crustacea: Isopoda) of Cos Sligo and Leitrim. *Irish Naturalists' Journal*, **25**: 273-277.

Cawley, M. (1997). *Armadillidium depressum* Brandt (Crustacea: Isopoda), new to Ireland. *Irish Naturalists' Journal*, **25**: 382.

Cawley, M. (2001). Notes and records on the Irish woodlice (Crustacea: Isopoda), including new sites for *Halophiloscia couchi* (Kinahan). *Bulletin of the Irish Biogeographical Society*, **25**: 211-217.

Chapman, P. (1993). *Caves and Cave Life*. New Naturalist Series, Harper Collins.

Chater, A.O. (1984). The distribution of *Metoponorthus cingendus*. *British Isopod Study Group Newsletter*, **No. 17**: 2-3.

Chater, A.O. (1986a). Woodlice in Ceredigion. *Dyfed Invertebrate Group Newsletter*, **No. 2**: 3-10.

Chater, A.O. (1986b). Recent woodlouse records from Dyfed. *British Isopod Study Group Newsletter*, **No. 21**: 4-5.

Chater, A.O. (1988). *Armadillidium pictum* in Radnorshire. *British Isopod Study Group Newsletter,* **No. 24**: 2.

Chater, A.O. (1989). Woodlice in Ceredigion. *Dyfed Invertebrate Group Newsletter,* **16**: 23.

Cole, G.A. & Minkley, W.L. (1968). A new species of aquatic isopod crustacean (Genus *Asellus*) from the Puebla Plateau, Central Mexico. *Proceedings of the Biological Society of Washington,* **81**: 755-760.

Collis, G.M. (2006). *Armadillidium vulgare* in West Central Scotland. *British Myriapod and Isopod Group Newsletter,* **No. 13**: 2.

Collis, G.M. (2007). Report on the 2006 BMIG meeting in Ayrshire. *Bulletin of the British Myriapod and Isopod Group,* **22**: 32-35.

Collis, G.M. & Collis, V.D. (2004). Woodlice on the Scottish islands of Bute, Islay and Mull. *Bulletin of the British Myriapod and Isopod Group,* **20**: 20-24.

Collis, G.M. & Harding, P.T. (2007). Charles Rawcliffe's discovery of the alien woodlouse *Styloniscus mauritiensis* (Barnard 1936). *Bulletin of the British Myriapod and Isopod Group,* **22**: 20-22.

Corbet, G.B. (1962). The 'Lusitanian Element' in the British Fauna. *Science Progress,* **50**: 177-191.

Cuppen, J.G.M., Glystra, R., van Beusekom, S., Budde, B.J. & Brock, T.C.M. (1995). Effects of nutrient loading and insecticide application on the ecology of *Elodea*-dominated freshwater microcosms. III: Responses of macroinvertebrate detritivores, breakdown of plant litter and final conclusions. *Archiv für Hydrobiologie,* **134**: 157-177.

Dandy, J.E. (1969). *Watsonian vice counties of Great Britain.* London, Ray Society.

Daws, J. (1991). Woodlice on Lundy. *British Isopod Study Group Newsletter,* **No. 31**: 2.

Daws, J. (1992a). *Porcellio laevis* – First Welsh records. *British Isopod Study Group Newsletter,* **No. 33**: 9.

Daws, J. (1992b). *Halophiloscia* in Kent. *British Isopod Study Group Newsletter,* **No. 33**: 9-10.

Daws, J. (1993a). BISG/BMG Field Meeting, Sussex. *British Isopod Study Group Newsletter,* **No. 35**: 1-3.

Daws, J. (1993b). Almost a new species for Britain. *British Isopod Study Group Newsletter,* **No. 36**: 4-5.

Daws, J. (1993c). The pirate BISG/BMG meeting in Lincolnshire. *British Isopod Study Group Newsletter,* **No. 36**: 5-6.

Daws, J. (1994a). *Leicestershire Woodlice.* Occasional Publications Series **9**. Leicestershire Entomological Society.

Daws, J. (1994b). Shetland Woodlice. *The Shetland Naturalist,* **1**: 82-84.

Daws, J. (1994c). *Porcellionides cingendus* in Suffolk. *British Isopod Study Group Newsletter,* **No. 37**: 3.

Daws, J. (1995a). The woodlouse *Stenophiloscia zosterae* – First record since 1977. *Suffolk Natural History* **31**: 25.

Daws, J. (1995b). *Trichoniscoides* in Suffolk. *British Isopod Study Group Newsletter,* **No. 38**: 2.

Daws, J. (1996). *Armadillidium pictum* in Leicestershire. *British Isopod Study Group Newsletter,* **No. 39**: 1.

Daws, J. (1997). The woodlouse *Ligidium hypnorum* in Leicestershire. *British Isopod Study Group Newsletter,* **No. 40**: 1.

Daws, J. (1999). Mineralised Woodlice. In: Connor, A. & Buckley, R. (eds) *Roman and Medieval Occupation in Causeway Lane, Leicester.* Leicester Archaeology Monographs **No. 5**.

Doogue, D. & Harding, P.T. (1982). *Distribution Atlas of Woodlice in Ireland.* An Foras Forbartha. Dublin.

Easter, J.K. (2005). A check-list, distribution and brief description of the terrestrial isopod species of Gibraltar. *Iberis,* **1**: 21-33.

Edney, E.B. (1954). *British woodlice.* Synopses of the British fauna, No. 9. London, Linnean Society.

Edney, E.B. (1953). The Woodlice of Great Britain and Ireland – a concise systematic monograph. *Proceedings of the Linnaean Society of London,* **164**: 49-98.

Enghoff, H. (1976). Morphological comparison of bisexual and parthenogenetic *Polyxenus lagurus* (Diplopoda: Polyxenida) in Denmark and Southern Sweden with notes on taxonomy, distribution and ecology. *Entomologiske Meddeldser,* **44**: 161-182.

Fiers, F. & Wouters, K. (1985). *Human impacts on the crustacean stygiofauna.* Proceedings of the Conférence – *Débat Invertébrés menaçants, Invertébrés menaces.* Gembloux, Faculte des Sciences Agronomiques de l'Etat.

Fowles, A. (1989). New applications of old techniques. *British Isopod Study Group Newsletter,* **No. 26**: 5.

Frankel, B., Sutton, S.L. & Fussey, G.D. (1981). The sex ratios of *Trichoniscus pusillus* Brandt (Crustacea: Oniscoidea). *Journal of Natural History,* **15**: 301-307.

Fussey, G.D. (1984). The distribution of the two forms of the woodlouse *Trichoniscus pusillus* Brandt (Isopoda: Oniscoidea) in the British Isles: A reassessment of geographic parthenogenesis. *Biological Journal of the Linnean Society,* **22**: 309-321.

Futter, S. (1998). Pill Woodlouse *Armadillidium vulgare* in Clydebank. *Glasgow Naturalist,* **2**: 62-63.

Gentile, G. & Sbordoni, V. (1998). Indirect methods to estimate gene flow in cave and surface populations of *Androniscus dentiger* (Isopoda: Oniscidea). *Evolution,* **52**: 432-442.

Gledhill, T., Sutcliffe, D.W. & Williams, W.D. (1993). British freshwater Crustacea Malacostraca: a key with ecological notes. *Freshwater Biological Association Scientific Publication* **No. 52**. Ambleside, Freshwater Biological Association.

Gorvett, H. & Taylor, J.C. (1960). A further note on tegumental glands in woodlice. *Proceedings of the Zoological Society of London,* **133**: 653-655.

Graça, M.A.S., Maltby, L. & Calow, P. (1993). Importance of fungi in the diet of *Gammarus pulex* and *Asellus aquaticus* I: feeding strategies. *Oecologia,* **93**: 139-144.

Gregory, S.J. (2001). Oxfordshire Woodlice: current status and distribution. *Bulletin of the British Myriapod and Isopod Group,* **17**: 60-75.

Gregory, S.J. (2002a). Report on the 2001 Field Meeting in Ireland: Woodlice. *Bulletin of the British Myriapod and Isopod Group,* **18**: 59-61.

Gregory, S.J. (2002b). Exclusive *Trichoniscoides. British Myriapod and Isopod Group Newsletter,* **No. 4**: 1.

Gregory, S.J. (2003). Non-marine Isopod Recording Scheme news. *British Myriapod and Isopod Group Newsletter,* **No. 6**: 2.

Gregory, S.J. (2004). BMIG expedition to Galicia – a preliminary woodlouse report. *British Myriapod and Isopod Group Newsletter,* **No. 9**: 2-3.

Gregory, S.J. (2005). *Porcellio spinicornis* in Western Scotland. *British Myriapod and Isopod Group Newsletter,* **No. 10**: 2-3.

Gregory, S.J. (2006). Recent changes to the names of British woodlice. *British Myriapod and Isopod Group Newsletter,* **No. 13**: 3-4.

Gregory, S.J. (2008). *Armadillidium pictum* Brandt, 1833 (Isopoda, Oniscidea) in Downton Gorge NNR, Herefordshire. *Bulletin of the British Myriapod and Isopod Group,* **23**: 13-14.

Gregory, S.J. (2009). A tropical woodlouse new to Britain. *British Myriapod and Isopod Group Newsletter,* **No. 18**: 1-2.

Gregory S.J. & Campbell J.M. (1995). *An Atlas of Oxfordshire Isopoda: Oniscidea (Woodlice).* Occasional Paper No. 17. Oxfordshire County Council.

Gregory, S.J. & Richards, P. (2008). Comparison of three often mis-identified species of pill-woodlouse *Armadillidium* (Isopoda: Oniscidea). *Bulletin of the British Myriapod and Isopod Group,* **23**: 9-12.

Gregory, S.J., Whiteley, D. & Wilde, I. (2001). Some observations of *Stenophiloscia zosterae* (Verhoeff, 1928) at Colne Point NNR, north Essex. *Bulletin of the British Myriapod and Isopod Group,* **17**: 79-80.

Gruner, H.-E. (1965). Krebstiere oder Crustacea. V. Isopoda. 1 lieferung. *Die Tierwelt Deutschlands,* **51**: 1-149. Jena.

Gruner, H.-E. (1966). Krebstiere oder Crustacea. V. Isopoda. 2 lieferung. *Die Tierwelt Deutschlands,* **53**: 151-380. Jena.

Gupta, S., Collier, J.S., Palmer-Felgate, A. & Potter, G. (2007). Catastrophic flooding origin of shelf valley systems in the English Channel. *Nature*, **448**: 342-345.

Hadley Centre website (2008): http://www.metoffice.gov.uk/research/hadleycentre/ (accessed April 2008).

Hames, C.A.C. (1987). Provisional atlas of the association between *Platyarthrus hoffmannseggi* and ants in Britain and Ireland. *Isopoda*, **1**: 9-20.

Harding, P.T. (1976a). *Provisional Atlas of the Crustacea of the British Isles, Part 1 Isopoda: Oniscoidea, Woodlice*. Institute of Terrestrial Ecology. Huntingdon.

Harding, P.T. (1976b). *Eluma purpurascens* Budde-Lund (Crustacea: Isopoda) a woodlouse new to Britain from Norfolk. *Transactions of the Norfolk & Norwich Naturalists' Society*, **23**: 267-268.

Harding, P.T. (1977). A re-examination of the work of W.E. Collinge on woodlice (Crustacea, Isopoda, Oniscoidea) from the British Isles. *Journal of the Society of the Bibliography of Natural History*, **8**: 286-315.

Harding, P.T. (1989). The occurrence of Asellidae in the British Isles. Part 1: *Asellus cavaticus*. *Isopoda*, **3**: 5-7.

Harding, P.T. (1990). An indexed bibliography of the distribution and ecology of woodlice (Crustacea, Oniscidea) in Great Britain (1830-1986). *Isopoda*, **4**: 1-32.

Harding, P.T. (2006). *Armadillidium pictum* at Tarren yr Esgob, Breconshire. *British Myriapod and Isopod Group Newsletter*, **No. 12**: 3.

Harding, P.T. & Collis, G.M. (2006). The occurrence of *Asellus communis* Say, 1818 (Crustacea, Isopoda) at Bolam Lake, Northumberland. *Bulletin of the British Myriapod and Isopod Group*, **21**: 8-11.

Harding P.T., Cotton M.J. & Rundle A.J. (1980). The occurrence of *Halophiloscia* (*Stenophiloscia*) *zosterae* Verhoeff, 1928 (Isopoda, Oniscidea) in Great Britain. *Crustaceana*, **39**: 111-112.

Harding, P.T. & Moon, H.P. (1976). A re-examination of *Asellus crypticus* Collinge, 1945 (Isopoda). *Crustaceana*, **30**: 109-110.

Harding, P.T. & Sutton, S.L. (1985). *Woodlice in Britain and Ireland: distribution and habitat*. Huntingdon, Institute of Terrestrial Ecology.

Harper, J. (2002). *Haplophthalmus montivagus* Verhoeff 1941 new to Wales. *Bulletin of the British Myriapod and Isopod Group*, **18**: 50-51.

Harper, J. (2004a). *Haplophthalmus montivagus* extended distribution. *Bulletin of the British Myriapod and Isopod Group*, **20**: 38-39.

Harper, J. (2004b). *Buddelundiella cataractae* inland in Wales. *Bulletin of the British Myriapod and Isopod Group*, **20**: 49.

Hidding, B., Michel, E., Natyaganova, A.V. & Sherbakov, D.Yu. (2003). Molecular evidence reveals polyphyletic origin and chromosomal speciation of Lake Baikal's endemic Asellid isopods. *Molecular Ecology*, **12**: 1509-1514.

Holdich, D.M. (1988). Wet Woodlice! *British Isopod Study Group Newsletter*, **No. 24**: 5-6.

Hopkin, S.P. (1986). *Armadillidium pulchellum*, editors note. *British Isopod Study Group Newsletter*, **No. 21**: 5.

Hopkin, S.P. (1987a). Biogeography of woodlice in Britain and Ireland. *Isopoda*, **1**: 21-36.

Hopkin, S.P. (1987b). *Metatrichoniscoides celticus* and *Trichoniscoides albidus*. *British Isopod Study Group Newsletter*, **No. 23**: 2.

Hopkin, S.P. (1989). *Ecophysiology of metals in terrestrial invertebrates*. Elsevier Applied Science. Barking.

Hopkin, S.P. (1990a). *Metatrichoniscoides leydigii* (Weber 1880). *British Isopod Study Group Newsletter*, **No. 28**: 1-2.

Hopkin, S.P. (1990b). *Trichoniscoides helveticus* Carl 1908. *British Isopod Study Group Newsletter*, **No. 28**: 2.

Hopkin, S.P. (1990c). *Trichoniscoides helveticus/sarsi*. *British Isopod Study Group Newsletter*, **No. 29**: 1.

Hopkin, S.P. (1991a). *A key to the woodlice of Britain and Ireland*. AIDGAP, Field Studies Council Publication No. 204. Preston Montford. (reprinted from *Field Studies* **7**: 599-650.)

Hopkin, S.P. (1991b). *Halophiloscia couchi* at St Bees. *British Isopod Study Group Newsletter*, **No. 32**: 2.

Hopkin, S.P., Hardisty, G. & Martin, M.H. (1986). The woodlouse *Porcellio scaber* as a 'biological indicator' of zinc, cadmium, lead and copper pollution. *Environmental Pollution*, **11**: 271-290.

Hopkin, S.P. & Roberts, A.W.P. (1987). A species of *Haplophthalmus* new to Britain. *Isopoda*, **1**: 37-48.

Hynes, H.B.N., Macan, T.T. & Williams, W.D. (1960). A key to the British species of Crustacea: Malacostraca occurring in fresh water. *Freshwater Biological Association Scientific Publication*, No 19. Ambleside: Freshwater Biological Association.

Irwin, T. (1982). Finding *Buddelundiella cataractae* Verhoeff. *British Isopod Study Group Newsletter*, **No. 15**: 6.

Irwin, A.G. (1992). *Metatrichoniscoides* sp. (Isopoda: Trichoniscidae), *Epipsocus lucifugus* (Psocoptera: Epipscocidae) and *Leptoiulus belgicus* (Diplopoda: Julida) new to Ireland and confirmation of *Rhynchodemus sylvaticus* (Tricladida: Rhynchodemidae) as an Irish species. *Irish Naturalists' Journal*, **24**: 106-110.

IUCN (2001). *IUCN Red List Categories and Criteria. Version 3.1*. Gland, Switzerland and Canbridge, UK: Species Survival Commission, IUCN.

Jefferson, G.T., Chapman, P., Carter, J. & Proudlove, G. (2004). The invertebrate fauna of the Ogof Ffynnon Ddu cave system, Powys, South Wales, UK. *Cave and Karst Science*, **31**: 63-76.

Joint Nature Conservation Committee (2008). *Common Standards Monitoring Guidance for Terrestrial and Freshwater Invertebrates*, Version March 2008. Joint Nature Conservation Committee.

Jones, G. (2008). *Trachelipus rathkii* (Brandt 1833); an Isopod new to Wales. *Bulletin of the British Myriapod and Isopod Group*, **23**: 17.

Jones, R.E. & Pratley, P. (1987). Woodlice of the Isles of Scilly. *Isopoda*, **1**: 49-54.

Kaushik, N.K. & Hynes, H.B.N. (1971). The fate of dead leaves that fall into streams. *Archiv für Hydrobiologie*, **68**: 465-515.

Kerney, M. (1999). *Atlas of the Land and Freshwater Molluscs of Britain and Ireland*. Colchester, Harley Books.

Knight, L. Undated [2007]. *Cave life in Britain*. Far Sawrey: Freshwater Biological Association.

Lambeck, K. (1996). Glaciation and sea-level change for Ireland and the Irish Sea since Late Devensian/Midlandian time. *Journal of the Geological Society, London*, **153**: 853-872.

Lee, P. (1993). Woodlice in Suffolk. *Transactions of the Suffolk Naturalists' Society*, **29**: 12-21.

Lee, P. (2003). *Porcellio spinicornis* in shingle. *British Myriapod and Isopod Group Newsletter*, **No. 7**: 4.

Lee, P., Barber, T. & Gregory, S. (2007). Collecting Centipedes, Millipedes and Woodlice: the true story. *Journal of the Amateur Entomologists' Society*, **66**: 117-123.

Magnin, E. & Leconte, O. (1971). Cycle vital d'*Asellus communis* sensu Racovitza (1920) (Crustacea, Isopoda) du lac Saint-Louis près de Montréal. *Canadian Journal of Zoology*, **49**: 647-655.

Magnin, E. & Leconte, O. (1973). Croissance relative chez le Crustacé isopode *Asellus communis* sensu Racovitza (1920) (Crustacea, Isopoda) du lac Saint-Louis près de Montréal. *Verhandlungen der Internationalen Vereinigung für Theoretische und Angewandte Limnologie*, **18**: 1488-1494.

Maltby, L. (1991). Pollution as a probe of life-history adaptation in *Asellus aquaticus* (Isopoda). *Oikos*, **61**: 11-18.

Martin, T.R. & Holdich, D.M. (1986). The acute lethal toxicity of heavy metals to peracarid crustaceans (with particular reference to freshwater asellids and gammarids). *Water Research*, **20**: 1137-1147.

Monahan, C. & Caffrey, J.M. (1996). The effects of weed control practices on the macroinvertebrate communities in Irish canals. *Hydrobiologia*, **340**: 205-211.

Moon, H.P. (1953). A re-examination of certain records for the genus *Asellus* (Isopoda) in the British Isles. *Proceedings of the Zoological Society of London*, **123**: 411-417.

Moon, H.P. & Harding, P.T. (1981). *A preliminary review of the occurrence of* Asellus *(Crustacea: Isopoda) in the British Isles*. Abbots Ripton: Biological Records Centre.

Moore, J.W. (1975). The role of algae in the diet of *Asellus aquaticus* L. and *Gammarus pulex* L. *Journal of Animal Ecology*, **44**: 719-730.

Morgan, I.K. (1994). An annotated list of the Woodlice of Carmarthenshire. *Dyfed Invertebrate Group Newsletter*, **No. 29**: 4-10.

Morgan, I.K. & Pryce R.D. Eds. (1995). *The Bulletin of the Llanelli Naturalists'*, **1**: 23.

Moseley, M. (1995). Comments on habitat selection by *Trichoniscoides saeroeensis*. *British Isopod Study Group Newsletter*, **No. 38**: 2.

Murphy, P.M. & Learner, M.A. (1982). The life history and production of *Asellus aquaticus* (Crustacea: Isopoda) in the River Ely, South Wales. *Freshwater Biology*, **12**: 435-444.

O'Meara, M. (2002). *The Woodlice of Waterford City & County: a checklist and atlas of the terrestrial isopods of Waterford at the start of the twenty-first century*. Fauna of Waterford Series, **No. 7**. Waterford.

Økland, J. & Økland, K.A. (1987). Asellen funnet I Stavanger. *Fauna*, **40**: 40-41.

Økland, K.A. (1980). *Ecology and distribution of Asellus aquaticus (L.) in Norway, including relation to acidification in lakes*. SNSF-project, Oslo-Ås, Norway, IR 52/80.

Oliver, P.G. (1983). The occurrence of *Buddelundiella cataractae* Verhoeff, 1930 (Isopoda, Oniscoidea) in Wales, Great Britain. *Crustaceana*, **44**: 105-108.

Oliver, P.G. & Meechan, C.J. (1993). *Woodlice*. Synopses of the British Fauna (New Series). Field Studies Council. Preston Montford.

Oliver, P.G. & Sutton, S. L. (1982). *Miktoniscus patiencei* Vandel, 1946. (Isopoda: Oniscoidea), a redescription with notes on its occurrence in Britain and Eire. *Journal of Natural History*, **16**: 201-208.

Oliver, P.G. & Trew, A. (1981). A new species of *Metatrichoniscoides* (Crustacea: Isopoda: Oniscoidea) from the coast of South Wales, U.K.. *Journal of Natural History*, **15**: 525-529.

Ormerod, S.J. & Walters, B. (1984). *Asellus cavaticus* Schiodte (Crustacea: Isopoda) from a hill stream in north Breconshire. *Nature in Wales*, **2**: 109.

Paoletti, M.G. & Hassal, M. (1999). Woodlice (Isopoda: Oniscidea): their potential for assessing sustainability and use as bioindicators. *Agriculture, Ecosystems & Environment*, **74**: 157-165.

Proudlove, G.S., Wood, P.J., Harding, P.T., Horne, D.J., Gledhill, T. & Knight, L.R.F.D. (2003). A review of the status and distribution of the subterranean aquatic Crustacea of Britain and Ireland. *Cave and Karst Science*, **30**: 53-74.

Rapp, W.F. (1988). *Trachelipus rathkii* in North America. *Isopoda*, **2**: 15-20.

Rawcliffe, C. (1987). Collecting in hothouses. *British Isopod Study Group Newsletter*, **No. 22**: 6.

Řezáč, M., Pekár, S. & Lubin, Y. (2008). How oniscophagous spiders overcome woodlouse armour. *Journal of Zoology* **275**: 64-71.

Richards, P. (1995). Millipedes, Centipedes and Woodlice of the Sheffield Area. *Sorby Record Special Series*, **No. 10**. Sheffield City Museum/Sorby Natural History Society.

Richards, P. (2004). Report on 2002 BMIG meeting in Derbyshire and south Yorkshire. *Bulletin of the British Myriapod and Isopod Group*, **20**: 42-48.

Richards, P. & Thomas, R. (1998). Woodlice and centipedes new to the region. *Sorby Record*, **34**: 78. Sorby Natural History Society, Sheffield.

Richardson, D.T. (1989). *Armadillidium pictum* Brandt in Yorkshire. *Isopoda*, **3**: 13-14.

Robinson, N.A. (2001). Observations on *Platyarthrus hoffmannseggi* and some other less common woodlice in NW England. *British Myriapod and Isopod Group Newsletter*, **No. 2**: 2.

Rossi, L. & Fano, A.E. (1979). Role of fungi in the trophic niche of the congenetic detritivorous *Asellus aquaticus* and *A. coxalis* (Isopoda). *Oikos*, **32**: 380-385.

Rossi, L. & Vitagliano-Tadini, G. (1978). Role of adult faeces in the nutrition of larvae of *Asellus aquaticus* (Isopoda). *Oikos*, **30**: 109-113.

Rundle, A. (1976). Kew Hothouse Woodlice. *British Isopod Study Group Newsletter*, **No. 10**: 2-3.

Schmalfuss, H. (1984). Eco-morphological Strategies in Terrestrial Isopods. In: Sutton, S.L. & Holdich, D.M. (eds). *The Biology of Terrestrial Isopods. Zoological Society of London Symposia*, **53**: 49-63.

Schmalfuss, H. (2004). World catalog of terrestrial isopods (Isopoda: Oniscidea). Available online at http://www.naturkundemuseum-bw.de/stuttgart/projekte/oniscidea-catalog/. An update of: Schmalfuss (2003) World catalog of terrestrial isopods (Isopoda: Oniscidea). *Stuttgarter Beitrage zur Naturkunde*, Serie A, **654**: 1-341.

Schotte M., Boyko C.B., Bruce N.L., Markham J., Poore G.C.B., Taiti S. & Wilson G.D.F. (eds) (2008, onwards). *World List of Marine Freshwater and Terrestrial Isopod Crustaceans*. Available online at http://www.marinespecies.org/isopoda.

Scott-Langley, D. (2002). Myriapoda (Chilopoda and Diplopoda) and Isopoda from the Isle of Mull and associated islands, Scotland. *Bulletin of the British Myriapod and Isopod Group*, **18**: 13-25.

Sheppard, E.M. (1968). *Trichoniscoides saeroeensis* Lohmander, an isopod new to the British fauna. *Transactions of the Cave Research Group of Great Britain*, **10**: 135-137.

Sunderland, K.D. & Sutton, S.L. (1980). A serological study of arthropod predation on woodlice in a dune grassland ecosystem. *Journal of Animal Ecology*, **49**: 987-1004.

Sutcliffe, D.W. (1972). Notes on the chemistry and fauna of water-bodies in Northumberland, with special emphasis on the distribution of *Gammarus pulex* (L.), *G. lacustris* Sars and *Asellus communis* Say (new to Britain). *Transactions of the Natural History Society of Northumberland*, **17**: 222-248.

Sutcliffe, D.W. (1974). Sodium regulation and adaptation to fresh water in the Isopod genus *Asellus*. *Journal of Experimental Biology*, **61**: 719-736.

Sutton, S.L. (1972). *Woodlice*. London, Ginn (reprinted 1980, Oxford, Pergamon Press).

Sutton, S.L. & Harding, P.T. (1989). Interpretation of the distribution of terrestrial isopods in the British Isles. In Ferrara, F., Argano, R., Manicastri, C., Schmalfuss, H. & Taiti, S. (eds) *Proceedings of the second symposium on the biology of terrestrial isopods. Monitore Zoologico Italiano (N.S.)*, **4**: 43-61.

Sutton, S.L., Harding, P.T. & Burn, D. (1972). *Key to British Woodlice*. London, Ginn.

Telfer, M.G. (2007). *Armadillidium pulchellum* (Zenker) new to East Anglia. *British Myriapod and Isopod Group Newsletter*, **No. 14**: 3.

Vandel, A. (1960). Isopodes Terrestres. Premiere partie. *Faune de France*, **64**. Paris.

Vandel, A. (1962). Isopodes Terrestres. Deuxieme partie. *Faune de France*, **66**. Paris.

Verovnik, R., Sket, B. & Trontelj, P. (2005). The colonization of Europe by the freshwater crustacean *Asellus aquaticus* (Crustacea: Isopoda) proceeded from ancient refugia and was directed by habitat connectivity. *Molecular Ecology*, **14**: 4355-4369.

Welter-Schultes, F.W. (2008). Bronze Age shipwreck snails from Turkey: first direct evidence for oversea carriage of land snails in antiquity. *Journal of Molluscan Studies*, **74**: 79-87.

Whitehead, P.F. (1988). New sites for *Trachelipus rathkei* in England. *Isopoda*, **2**: 11-14.

Whitehouse, N.J. (2006). The Holocene British and Irish ancient forest fossil beetle fauna: implications for forest history, biodiversity and faunal colonization. *Quaternary Science Reviews*, **25**: 1755-1789.

Wickenberg, M. & Reynolds, J.D. (2002). A recent Irish record of the woodlouse *Acaeroplastes melanurus* (Budde-Lund, 1885) (Isopoda: Porcellionidae), considered to be extinct in the British Isles. *Bulletin of the Irish Biogeographical Society*, **26**: 60-63.

Wieser, W. (1984). Ecophysiological adaptations of terrestrial isopods: a brief review. In: Sutton, S.L. & Holdich, D.M. (eds). *The Biology of Terrestrial Isopods. Zoological Society of London Symposia*, **53**: 49-63.

Wijnhoven, H. (2001a). Landpissebedden van de ooijpolder: deel 2. ecologie (Crustacea: Isopoda: Oniscidea). *Nederlandse faunistische mededelingen*, **14**: 23-78.

Wijnhoven, H. (2001b). Biologie en ecologie van de Nederlandse pissebedvliegen (Diptera: Rhinophoridae). *Nederlandse faunistische mededelingen*, **15**: 91-109.

Wijnhoven, H. & Berg, M.P. (1999). Some notes on the distribution and ecology of iridovirus (Iridovirus, Iridoviridae) in terrestrial isopods (Isopoda, Oniscidae). *Crustaceana*, **72**: 145-156.

Williams, T. & Franks, N. (1985). Research on *Platyarthrus hoffmannseggii* at Bath University. *British Isopod Study Group Newsletter*, **No 19**: 6-7.

Williams, W.D. (1962). Notes on the ecological similarities of *Asellus aquaticus* (L.) and *A. meridianus* Rac. (Crustacea, Isopoda). *Hydrobiologia*, **20**: 1-30.

Williams, W.D. (1970). A revision of the North American epigean species of *Asellus* (Crustacea: Isopoda). *Smithsonian Contributions to Zoology*, **49**. Washington: Smithsonian Institution.

Williams, W.D. (1972). Occurrence in Britain of *Asellus communis* Say, 1818, a North American freshwater isopod. *Crustaceana Supplement*, **3**: 134-138.

Williams, W.D. (1979). The distribution of *Asellus aquaticus* and *A. meridianus* (Crustacea, Isopoda) in Britain. *Freshwater Biology*, **9**: 491-501.

Index to species and families

Synonyms are given in *italics*. Start of main sections and maps are given in **bold**, illustrations in *italics*.